INTERNATIONAL CENTRE FOR MECHANICAL SCIENCES

COURSES AND LECTURES - No. 265

MULTI-USER COMMUNICATION SYSTEMS

EDITED BY
G. LONGO
UNIVERSITA' DI TRIESTE

Springer-Verlag Wien GmbH

Originally published by CISM, Udine in 1981.

ISBN 978-3-211-81612-7 ISBN 978-3-7091-2900-5 (eBook)
DOI 10.1007/978-3-7091-2900-5

PREFACE

Here is a set of qualified contributions to one of the most fascinating and intriguing problems in an area between system theory and communication theory, which is rapidly developing today, namely multiterminal or distributed communication.

I am grateful to the lecturers who accepted my invitation and came to Udine in July 1979, to talk about this topic and to illustrate its numerous facets. I am also grateful to the audience, who contributed in an active way to the success of the school.

The delay with which this volume is published although being unpleasant in itself, gave the authors the possibility of updating their contributions, thus making the written text even more valuable than the set of lectures from which it developed.

Just a few days before this book was printed, the Rector of CISM, professor Wacław Olszak, passed away. It is with great sorrow that I announce his death to the friends of our Centre, and I wish to devote this book to him. He had been one of the founders of CISM and had spent the last twelve years of his life in building and extending the activity of this Centre. We all are greatly indebted to him.

Giuseppe Longo

Udine, December 1980

CONTENTS

Page

The Poisson Multiple-Access Conflict Resolution Problem
by T. Berger

Preface 1
1. Description of the Problem 2
2. Reformation-Fishing in a Poisson Stream with a Weak Net 3
3. Slotted ALOHA 4
4. The 0.487 Algorithm 9
5. Upper Bounds 18
References 26

Performance Analysis of Multi-User Communication Systems: Access Methods and Protocols
by E. Gelenbe

1. Introduction 29
2. Random Access to a Single Broadcast Channel 30
References 47

About a Combinatorial Proof of the Noisy Channel Coding Theorem
by J. Körner

Introduction 49
Statement of the Problem 50
Existence of Good Codes 57
Converse Results 63
Universal Coding 68
Outlook 70
References 71

Collision-Resolution Algorithms and Random-Access Communications
by J.L. Massey

1. Introduction 73
2. Collision-Resolution Algorithms 76
3. Traffic-Independent Properties of the Capetanakis Collision-Resolution Algorithm . . 83
4. Random-Access via Collision-Resolution 95
5. Effects of Propagation Delays, Channel Errors, etc. 121
6. Summary and Historical Credits 134
Acknowledgement 135
Postscript 135
References 137

Spread-Spectrum Multiple-Access Communications
by M.B. Pursley

I. Spread-Spectrum Communications: a Random Signal Model 141
II. Introduction to Binary Direct-Sequence SSMA Communications 146
III. Multiple-Access Interference 154
IV. Properties of the Continuous-Time Crosscorrelation Functions 164
V. Quaternary Direct-Sequence SSMA Communications 181
Acknowledgements 196
References and Selected Bibliography 197

Some Tools for the Study of Channel-Sharing Algorithms
by G. Ruget

I. About the Algorithms 201
II. Destabilization of ALOHA: Introduction to Large Deviations 210
3. Large Deviators (continued) 219
4. Fast Simulations (continued) 225
Bibliography 230

Asymmetric Broadcast Channels
by A. Sgarro

1. Introduction 233
2. Preliminaries 235

3. Definitions 236
4. Results 238
References 241

A Linear Programming Bound for Codes in a 2-Access Binary Erasure Channel
by H.C.A. van Tilborg

I. Introduction 244
II. Definitions and Lemmas 245
III. An Upper Bound 248
IV. Uniquely Decodable Codes 253
V. Some Examples for $d_L = 4$ 255
References 257

THE POISSON MULTIPLE-ACCESS CONFLICT RESOLUTION PROBLEM

TOBY BERGER
School of Electrical Engineering
Cornell University, Ithaca, NY 14853

PREFACE

The major thrust of system theory research in recent years has been directed at multiterminal, or decentralized, problems. These are characterized by the fact that system functions – such as coding, routing, control, decision, and estimation – are effected simultaneously at several physically distinct sites within the system. The information available differs, in general, from site to site. This necessitates the development of distributed algorithms aimed at cooperatively achieving optimum, or near-optimum, performance in the face of topologically complex information patterns. In previous lectures at CISM, several investigators including myself have treated multiterminal information theory problems involving distributed algorithms for encoding and decoding. In these lectures I concentrated instead on an intriguing problem in multiterminal communication theory, the conflict resolution problem in packet-switched communication networks. After stating the problem, we recast it as one of "fishing in a Poisson stream with a weak net". The slotted ALOHA protocol is described ana analyzed. Gallager's improvement on Capetanakis's algorithm is then treated in considerable detail; a closed-form expression for its efficiency is derived. Recursive equations developed by the Russian and French schools are presented as an alternative means of analyzing this algorithm and others. Then we present Pippenger's and Molle's upper bounds to the maximum efficiency attainable. We conclude with what we feel to be a convincing case for our belief that Molle's method can be improved upon to yield an upper bound of 0.5254 which is only 0.0377 above the efficiency of 0.4877 achieved by the optimized version of Gallager's algorithm.

The final text of these lectures was compiled in April 1980. In this regard reference is made to some results of recent vintage obtained since the lectures themselves were delivered in July 1979. I am particularly grateful for permission to discuss the interesting work of M. Molle prior to its publication.

The Poisson Multiple-Access Conflict Resolution Problem

1. Description of the Problem

In multiple-access packet communication systems, information generated at many different sites must be cooperatively multiplexed onto a common communication link. Each site generates short bursts of information at random instants. Each such burst is encoded into the form of a fixed-length packet. Said packets can be transmitted only during so-called "slots" of duration equal to the packet length. Wherever two or more packets compete for the same slot, a collision occurs. In the event of a collision, none of the messages that collided gets sent and notice of the occurrence of the collision is broadcasted to all sites. We shall consider here the case in which there is no way of distinguishing the sites; i.e., they have no names or numbers. This is not the most important case in practice, but it raises challenging and intriguing theoretical questions.

A cooperative procedure, or protocol, must be adopted for retransmitting in subsequent slots those packets that have suffered collisons. The objective is to design this protocol so as to maximize the fraction of slots devoted to exactly one message (i.e., neither empty nor wasted because of a collision). The general protocol, or multiplexing algorithm, has the following form. On the basis of the past history of empty slots, utilized slots, and slots with collisions, a subset of the time axis is selected. Any packet yet to be successfully transmitted that encodes a message originally generated at an instant belonging to said subset tries for the next slot. In the event of a collision, a smaller time set is specified next time, and so on, until the collision is resolved.

Let N(t) denote the random number of messages generated up to time t by all the sites combined. We shall assume that $\{N(t), t \geq 0\}$ is a homogeneous Poisson process of intensity λ. Let Δ_n denote the random delay between the instant at which the nth message is generated and the instant at which its packet is successfully transmitted. Given any protocol, let $\bar{\lambda}$ denote the supremum of all the values of λ for which $\limsup_{n\to\infty} E\Delta_n$ is finite, where E denotes statistical expectation. Let λ_{max} denote the supremum of $\bar{\lambda}$ over all protocols. The problem we wish to solve is to find λ_{max} and to exhibit protocols for which $\bar{\lambda}$ is arbitrarily close to λ_{max}.

If time is measured in units equal to the packet duration (slot duration), then it is known that $0.4877 \leq \lambda_{max}$. We establish this lower bound in Section 4 by devising an explicit formula for $\bar{\lambda}$ for a particular protocol originally suggested by Gallager.[1] In section 5.1 we present an information-theoretic argument of Pippenger[2] which shows that $\lambda_{max} \leq 0.7448$. Section 5.2 is devoted to explication of a "genie" argument of Molle[3] which shows that $\lambda_{max} \leq 0.6731$. In Section 5.3 we suggest how Molle's approach might be extended to prove that $\lambda_{max} \leq 0.5254$.

2. Reformulation - Fishing in a Poisson Stream with a Weak Net

The problem described in Section 1 may be reformulated as one of "fishing in a Poisson stream with a weak net". Specifically, imagine that ideal point fish each weighting 1 pound are distributed in a one-dimensional homogeneous Poisson stream. At one-minute intervals we may dam off a segment of the stream of arbitrary length and deploy a fishing net therein. The net is capable of supporting 1.5 pounds and hence of catching one fish. If two or more fish are in the segment, we say it is congested. In such an instance the net temporarily opens, the fish fall back into the stream between the barriers, and the only information we receive is that at least two fish still reside between the barriers in question. The objective is to device a sequential fishing strategy that maximizes the "effishiency" η defined by

$$\eta \triangleq \liminf_{n\to\infty} n^{-1} \text{ (expected number of fish caught in n minutes).}$$

We seek η_{max}, the supremum of η over all fishing strategies, and schemes that come close to achieving it. (If the fishing analogy does not appeal to you, simply view the problem as one of isolating the events in a Poisson stream as efficiently as possible when told at each step whether the set you have selected contains 0, 1 or more than 1 event).

It is easy to see that $\eta_{max} \leq \lambda_{max}$. It is not hard to show that, if the net can be deployed over a union of several dammed-off segments whenever this is desired, then $\eta_{max} = \lambda_{max}$. It is believed that η_{max} is not affected if one is restricted to fishing in single segments (intervals), but this remains to be proven.

3. Slotted ALOHA

In order both to provide historical perspective and to develop appreciation for the improvements afforded by certain protocols devised recently, we shall begin with a discussion of the so-called slotted ALOHA algorithm. This algorithm, introduced by Roberts[4] in 1972, was incorporated into the ALOHA computer communication net operated by the University of Hawaii.

Suppose we have M message generation sites the mth of which generates a random number $V_m(k)$ of new packets between $t = k-1$ and $t = k$. For fixed m we assume that the $X_m(k)$, k = 1, 2, ..., are independent and identically distributed random variables. Then the average number of messages per second generated at site m is

$$\lambda_m = E\, V_m(k)$$

which does not depend on k. Assume that the message generation processes $\{V_m(k)\}$, $1 \leq m \leq M$, are mutually independent.

If slot conflicts can be resolved efficiently enough that with probability 1 the number $X_m(k)$ of packets queued for transmission at site m does not diverge as $k\to\infty$ for any m, then a steady state will be reached in which all messages generated eventually are transmitted with finite average delay. In such an instance the overall system throughput is

$$\lambda = \sum_{m=1}^{M} \lambda_m \quad .$$

It is of interest to consider the limit as $M\to\infty$ and all the $\lambda_m\to 0$ in such a way that $\lambda_1 + \ldots + \lambda_m$ remains fixed at λ. If throughput value λ can be achieved by some protocol in this limit, then λ clearly is a lower bound to λ_{max} as defined in Section 1.

The slotted ALOHA protocol resolves conflicts via a random retransmission scheme. If $X_m(k) \geq 1$ (i.e., if there is a queue of one or more packets at site m at t = k), then with probability p_m an attempt is made to transmit the packet at the head of the queue during the kth slot, $k \leq t < k + 1$. We shall show below that, for any finite M, it is possible to match the transmission probabilities p_m to the message generation rates λ_m such that any overall throughput $\lambda < e^{-1}$ is attainable. In the limit as $M\to\infty$ and $\lambda_m\to 0$, it unfortunately becomes necessary to send $p_m\to 0$

in order to achieve a positive overall throughput. Since the average delay each packet suffers before being successfully transmitted obviously diverges as $p_m \to 0$, the slotted ALOHA result does not allow us to conclude that $\lambda_{max} \geq e^{-1}$. However, by using an adaptively time-varying transmission probability $p_m(k)$, we can overcome the instability problems associated with slotted ALOHA and deduce that $\lambda_{max} \geq e^{-1}$.

Let us consider first the case of constant retransmission probabilities $p_m(k) = p_m$ for all k. In this case the multidimensional queue $\underline{X}(k) = (X_1(k), \ldots, X_M(k))$ is a homogeneous vector Markov chain. If the expected change of $X_m(k)$ over one slot time, namely

$$\Delta_m(k) \triangleq E[X_m(k+1) - X_m(k)] ,$$

is negative for each value of m in the limit as $k\to\infty$, then the Markov chain is ergodic. This means that the same long run statistics will prevail regardless of the initial state $\underline{X}(0)$ and, in particular, that there is zero probability that unbounded queue lengths will occur at any of the sites. We proceed to derive a sufficient condition for ergodicity of the slotted ALOHA protocol.

During the kth slot the size of the mth queue will be increased by $V_m(k)$ and will be decreased by 1 if $X_m(k) \geq 1$ and site m transmits successfully during the kth slot. Thus,

$$\begin{aligned}\Delta_m(k) &= E\,V_m(k) - P[X_m(k) \geq 1 \text{ and site m transmits successfully}]\\ &= \lambda_m - P[X_m(k) \geq 1]\; p_m \prod_{i\in A_m(k)} (1 - p_i)\end{aligned}$$

where $A_m(k) = \{i \neq m: X_i(k) \geq 1\}$ is the set of all sites other than m which have a queue at time k. If $\lim_{k\to\infty} P[X_m(k) \geq 1] < 1$, then $\lim_{k\to\infty} \Delta_m(k)$

cannot possibly be positive. This is because a positive limiting value for $\Delta_m(k)$ implies that $X_m(k)$ diverges as $k\to\infty$ and hence that $\lim_{k\to\infty} P[X_m(k) = 0]$ cannot possibly be zero. Therefore, we need consider only the case $\lim_{k\to\infty} P(X_m(k) \geq 1] = 1$. In such a case we get

$$\lim_{k\to\infty} \Delta_m(k) = \lambda_m - p_m \lim_{k\to\infty} \prod_{i\varepsilon A_k(m)} (1 - p_i) \leq \lambda_m - p_m \prod_{i\neq m} (1 - p_i)$$

A sufficient condition for $\lim_{k\to\infty} \Delta_m(k) < 0$, and hence for ergodicity, is seen to be

$$\lambda_m < p_m \prod_{i\neq m} (1 - p_i) \stackrel{\Delta}{=} Q_m \quad \text{for each m.}$$

It can be shown that, if $\lambda_m > Q_m$ for all m, then the Markov chain (system of queues) is non-recurrent; that is, all the queues surely diverge. If $\lambda_m \geq Q_m$ for some but not all values of m, we have a borderline case in which the chain may be recurrent but not ergodic.

To see that any $\lambda < e^{-1}$ is achievable, choose

$$p_m = e\,\lambda_m (1 + e\,\lambda_m)^{-1} .$$

Then the ergodicity condition $\lambda_m < Q_m$ reads

$$\lambda_m < \left(\frac{e\,\lambda_m}{1 + e\,\lambda_m}\right) \prod_{i\neq m} \left(1 - \frac{e\,\lambda_i}{1 + e\,\lambda_i}\right) = e\,\lambda_m \prod_{i=1}^{M} \left(\frac{1}{1 + e\,\lambda_i}\right)$$

or

$$\prod_{i=1}^{M} (1 + e\,\lambda_i) < e \quad .$$

From the fact that the geometric mean of a set of positive numbers always is less than the arithmetic mean, we deduce that

$$\prod_{i=1}^{M} (1 + e\,\lambda_i) \leq (\frac{1}{M} \sum_{i=1}^{M} 1 + e\,\lambda_i)^M = (1 + \frac{e\,\lambda}{M})^M \leq e^{e\,\lambda}$$

where the last inequality is a consequence of the fact that $(1 + x/n)^n$ is an increasing function of n with limit e^x. Accordingly, the system is ergodic if $e^{e\,\lambda} < e$, i.e., if $\lambda < e^{-1}$ as was to be shown.

Let us consider the limit as the number M of sites tends to infinity in such a way that the total expected traffic generated per second remains fixed at λ by virtue of the individual generation rates λ_m tending to zero. For simplicity, let $\lambda_m = \lambda/M$ and let $p_m = p$. If p remains constant in the limit as $M\to\infty$, then the ergodicity inequality reads

$$\lambda_m = \frac{\lambda}{M} < p(1 - p)^M = Q_m$$

Since $Mp(1 - p)^M \to 0$ as $M\to\infty$, this inequality fails to hold in the limit for any $\lambda>0$. In fact, the reverse inequality holds which implies the system is nonrecurrent (unstable). The only way to prevent the inequality from being violated in the limit is to let $p\to 0$ as $M\to\infty$ in such a way that $Mp(1 - p)^{M-1} \to$ const. $> \lambda$. Although this is possible, letting $p\to 0$ implies that the average waiting time of a message tends to infinity, a thoroughly unacceptable situation. Hence, we conclude that slotted ALOHA breaks down in the limit as $M\to\infty$ by virtue of total degeneration of throughput and/or delay.

It is possible to modify the ALOHA algorithm to avoid instability in the limit as $M\to\infty$. One way of doing this is to employ a time-dependent retransmission probability. For simplicity assume that the sites are statistically identical. Let $\{p_s\}$ be a decreasing sequence of probabilities that tend to 0 and $s\to\infty$. At time k all active sites (i.e., sites with $X_m(k) \geq 1$) send a message with probability $p(k) = p_s$, where s with varies with k according to the following prescription:

$$p(k+1) = \begin{cases} p_{s+1} & \text{if a conflict occurs at time } k \\ p_s & \text{if successful transmission occurs at time } k \\ p_{s-1} & \text{if } s \geq 1 \text{ and no one tries to send at time } k \end{cases}$$

This yields a distributed algorithm that adapts to the local traffic density. With this modification slotted ALOHA becomes a viable technique for practical situations in which the number M of users is a slowly varying function of time with a large dynamic range. The stability properties of this modified version of slotted ALOHA depend on the quantity $\lim_{s\to\infty} sp_s$ in a complicated manner that has been studied by Mikhailov.[5] We do not discuss that dependence here because in Section 4 we provide simpler means of obtaining better lower bounds to λ_{max}.

4. The 0.487 Algorithm

In the terminology of Section 2, we describe an "interval-fishing" strategy for which $\eta > 0.487$. The description and analysis are new, but the algorithm is Gallager's[1] improved version of a strategy devised by Capetanakis.[6]

The "neighbor" of interval I is defined to be the interval adjacent to I on the right and of the same length as I unless a different length is specified. Let x be a length parameter to be optimized subsequently. The states shown in Figure 1 have the following significances.

State x - fish next in the neighbor of length x

State 1/2 - fish next in the left half of the interval just fished

State N - fish next in the neighbor

State N/2 - fish next in the left half of the neighbor.

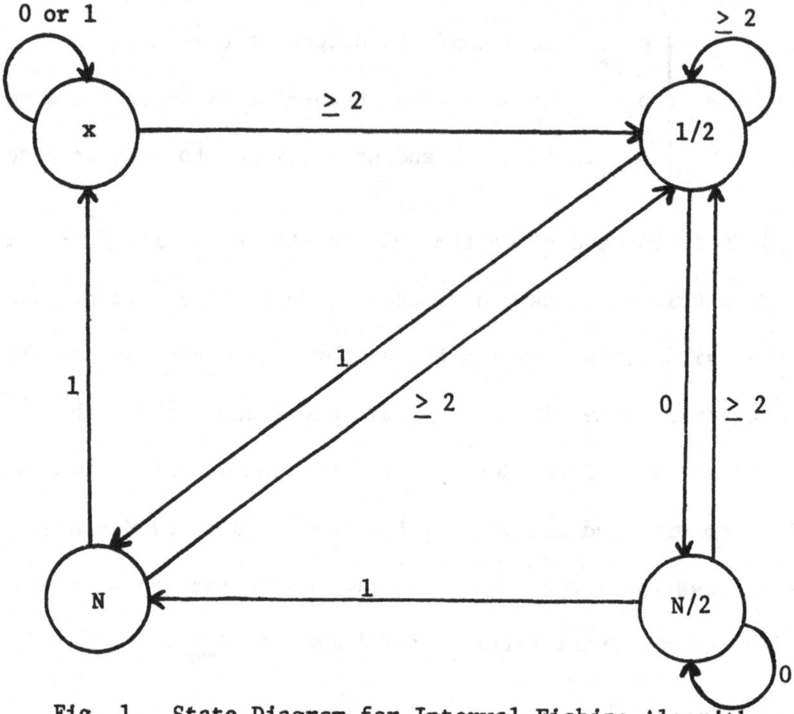

Fig. 1. State Diagram for Interval-Fishing Algorithm

The algorithm is initiated in state x by fishing first in [0,x]. The state transitions are indicated by arrows labeled by our knowledge of how many fish were in the interval that we just fished. No arrow labeled "0" emerges from state N because the interval one explores upon entering state N always harbors at least one fish.

By using the fact that the numbers of fish in non-overlapping Poisson stream segments are independent, it is not difficult to show that state x is a Markov state; the other states are not. It also can be seen that the leftmost fish always is the next one to be caught, which corresponds to "first in, first out" (FIFO) behavior of the associated packet conflict resolution protocol.

We shall now determine W(x), the average number of states visited between successive returns to state x, and C(x), the average number of fish caught between successive returns to state x. The "effishiency" of the algorithm is*

$$\eta(x) = C(x)/W(x) \ ,$$

which we subsequently maximize over x.

Every interval fished is of length $x2^{-k}$ for some value of $k \in \{0, 1, 2, \ldots\}$ so

$$W(x) = \sum_{k=0}^{\infty} \sum_{i=1}^{2^k} p_{ik}(x)$$

where $p_{ik}(x)$ is the probability that we fish in the interval $I_{ik} \triangleq [(i-1) \times 2^{-k}, i \times 2^{-k}]$ before the first return to state x. Assume without loss of generality that the Poisson stream has intensity 1; for a general intensity γ, simply replace x by γx throughout. We have

$$f_o \triangleq \Pr(2 \text{ or more fish in } [0,x]) = 1-(1 + x)\, e^{-x}$$

Toward evaluating p_{ik}, we introduce the symbols p_k, q_k, r_k, s_k and t_k for $k \geq 1$ via the following definitions wherein a k-interval is any interval of length $\ell_k = \ell_k(x) = x2^{-k}$

$p_k \triangleq$ Pr[congestion in the left half of a (k-1)-interval known to be congested]

*The Markovianness of state x allows us, via a law of large numbers argument, to express the long-run efficiency η as the ratio of the expected number of fish caught between successive visits to state x to the expected time it takes to catch them, despite the fact that η is not defined this way.

$$= [1-(1+\ell_k)e^{-\ell_k}]/[1-(1+\ell_{k-1})e^{-\ell_{k-1}}] \triangleq f_k/f_{k-1}$$

$q_k \triangleq$ Pr[no fish in left half of a congested (k-1)-interval]

$r_k \triangleq$ Pr[exactly one fish in left half of a congested (k-1)-interval]

$s_k \triangleq$ Pr[congestion in a k-interval known to contain at least one fish]

$t_k \triangleq q_k + r_k s_k$

Given a typical (k-1)-interval, let L and R denote the numbers of fish in its left and right halves, respectively. Then

$$r_k = P[L = 1|L + R \geq 2) = P(L = 1,\ L + R \geq 2)/P(L + R \geq 2) \,.$$

$$= P(L = 1,\ R \geq 1)/f_{k-1} = P(L = 1)\ P(R \geq 1)/f_{k-1}$$
$$= \ell_k e^{-\ell_k}(1-e^{-\ell_k})/f_{k-1}$$

$$q_k = P(L = 0|L + R \geq 2) = P(L = 0)\ P(R \geq 2)/f_{k-1} = e^{-\ell_k}\ f_k/f_{k-1}$$

$$s_k = P(R \geq 2|R \geq 1) = P(R \geq 2)/P(R \geq 1) = f_k/(1 - e^{-\ell_k})$$

$$t_k = q_k + r_k s_k = (e^{-\ell_k} + \ell_k e^{-\ell_k})\ f_k/f_{k-1}$$

$$p_k + t_k = [1 + (1 + \ell_k)e^{-\ell_k}]\ f_k/f_{k-1}$$

The values of the p_{ik} for $1 \leq i \leq 2^k$ for $0 \leq k \leq 3$ are shown in terms of the p_k, q_k, r_k, s_k and t_k in Figure 2 above the corresponding intervals I_{ik}. The general pattern for expressing the p_{ik} in terms of the p_j, r_j and t_j is evident from Figure 2.

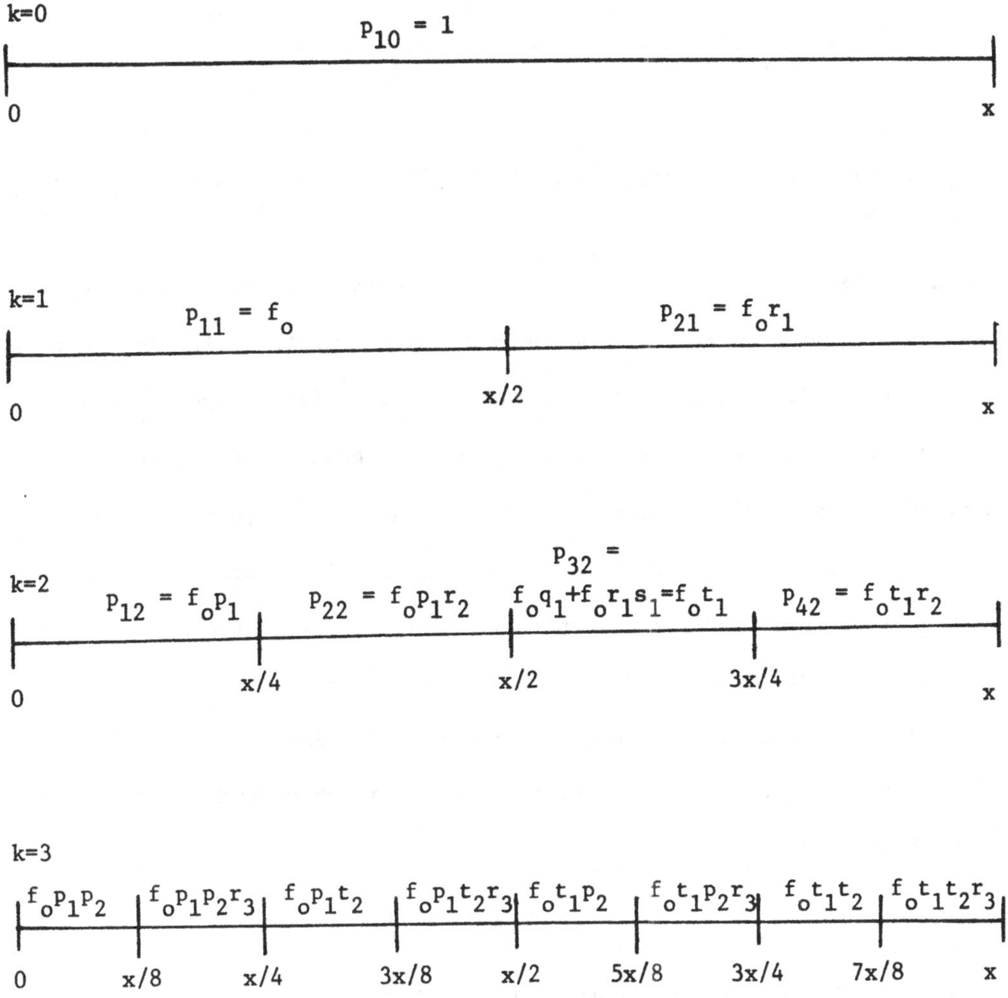

Figure 2. The p_{ik} for $1 \leq i \leq 2^k$ for $0 \leq k \leq 3$

We deduce that

$$W(x) = 1 + f_o(1 + r_1) + f_o(p_1 + t_1)(1 + r_2) + f_o(p_1 + t_1)$$

$$(p_2 + t_2)(1 + r_3) + \ldots = 1 + f_o \sum_{k=0}^{\infty} (1 + r_{k+1}) \prod_{j=1}^{k} (p_j + t_j) ,$$

where the empty product when k=0 is interpreted as 1.

It follows that

$$W(x) = 1 + \sum_{k=0}^{\infty} [f_k + \ell_{k+1} e^{-\ell_{k+1}} (1-e^{-\ell_{k+1}})] \prod_{j=1}^{k} [1+(1+\ell_j)e^{-\ell_j}] ,$$

or

$$W(x)=1+ \sum_{k=0}^{\infty} [1-(1+3x2^{-(k+1)})e^{-x2^{-k}} + x2^{-(k+1)} e^{-x2^{-(k+1)}}] \prod_{j=1}^{k} [1+(1+x2^{-j})e^{-x2^{-j}}]$$

The above exact expression for $W(x)$, albeit cumbersome in appearance, is readily evaluated on a programmable desk calculator to arbitrary accuracy.*

We now calculate $C(x)$, the average number of fish caught between successive returns to state x. Note that $C(x)$ equals the difference between the expected number of fish in the entirety of $[0,x]$ and the expected number in the portion of $[0,x]$ that still is unresolved as to fish content when we first return to state x. Study of the algorithm reveals that this portion is always a single interval of the form $(b,x]$. The aforementioned Markovianness assures that the interval $(b,x]$ is "fresh water," (i.e., statistically indistinguishable from the unexplored water beyond x), so

$$C(x) = x - E(x-b) = Eb.$$

Next note with the aid of Figures 1 and 2 that

$$x-b = \sum_{k=1}^{\infty} x2^{-k} \varepsilon_k$$

*Indeed, it is easy to bound the sum of all the terms for $k>n$ by a simple function of n and x that decays exponentially rapidly in n.

where

$$\varepsilon_k = \begin{cases} 1 & \text{if we ever fish in a congested k-interval when in state 1/2 or state N/2 (since the neighbor of said k-interval then will remain unresolved)} \\ 0 & \text{otherwise} \end{cases}$$

Thus,

$$E(x-b) = x \sum_{k=1}^{\infty} 2^{-k} \Pr(\varepsilon_k=1)] = x \sum_{k=1}^{\infty} 2^{-k} f_0 p_k \prod_{j=1}^{k-1} (p_j + t_j)$$

$$= x \sum_{k=1}^{\infty} 2^{-k} f_0 (f_k/f_{k-1}) \prod_{j=1}^{k-1} [1+(1+\ell_j)e^{-\ell_j}](f_j/f_{j-1})$$

$$= x \sum_{k=1}^{\infty} 2^{-k} f_k \prod_{j=1}^{k-1} [1+(1+\ell_j)e^{-\ell_j}]$$

$$= x \sum_{k=1}^{\infty} 2^{-k}[1-(1+x2^{-k})e^{-x2^{-k}}] \prod_{j=1}^{k-1} [1+(1+x2^{-j})e^{-x2^{-j}}]$$

It follows that the "effishiency" as a function of x is

$$\eta(x) = \frac{x\{1 - \sum_{k=1}^{\infty} 2^{-k}[1-(1+x2^{-k})e^{-x2^{-k}}] \prod_{j=1}^{k-1} [1+(1+x2^{-j})e^{-x2^{-j}}]\}}{1+\sum_{k=0}^{\infty} [1-(1+3x2^{-(k+1)})e^{-x2^{-k}}+x2^{-(k+1)}e^{-x2^{-(k+1)}}] \prod_{j=1}^{k} [1+(1+x2^{-j})e^{-x2^{-j}}]}$$

Computations reveal that $\eta(x)$ has a maximum of 0.4871 at $x = 1.2661$. This establishes the lower bound $0.487 < \eta_{max}$.

Gallager's algorithm, which we shall also refer to as the "0.487 algorithm," was discovered apparently independently not only by Gallager but also by Ruget[7] in Paris and by Tsybakov <u>et al.</u>[8] in Moscow. Their method of analyzing the algorithm's performance differs substantially from that presented above. It is of sufficient interest to merit presentation here because it readily adapts to situations in which conflicted intervals are not divided exactly in half and to certain considerably more complicated algorithms. The approach consists of developing a system of recurrence equations for the quantities

$W_k \triangleq$ average amount of work needed to resolve a conflict between k messages up to the point at which two successive slots have successful transmission.

Of course, in practice one knows only that $k \geq 2$, not what k actually is, so the recurrence relations are purely an analytical tool. (Pippenger[2] has shown that, if the value of k actually were known, then efficiencies arbitrarily close to 1 could be achieved).

Let $P_{k,i}$, $0 \leq i \leq k$, be the probability that i out of the k users involved in the conflict attempt to retransmit in the next slot. (This randomness of retransmission can be achieved either by a probabilistic retransmission scheme, as in slotted ALOHA and variants thereof, or by deterministically narrowing the time interval in which one's message must have been generated in order for one to be allowed to try again to send it.) We obtain for the 0.487 algorithm

$$W_k = P_{k,0}(1+W_k) + P_{k,k}(1+W_k) + P_{k,1}(2+W_{k-1}) + \sum_{i=2}^{k-1} P_{k,i}(1+W_i)$$

with boundary conditions $W_0 = W_1 = 1$. Rearranging this yields

$$(1-P_{k,0}-P_{k,k})\, W_k = 1 + P_{k,1}(1+W_{k-1}) + \sum_{i=2}^{k-1} P_{k,i}(1+W_i)$$

This is clearly a simple recursive formula for W_k in terms of earlier W_i, so we can generate the W_k by a simple computational procedure. The overall work thus is

$$W = \sum_{k=0}^{\infty} q_k W_k$$

where q_k is the probability that k messages lie in an unexplored basic interval of duration x. These are simply the Poisson probabilities $q_k = x^k e^{-x}/k!$

A similar recursive procedure can be used to determine the expected duration of the interval length that gets explored before the first time two fish in a row are caught. Let

$$B_k = (\frac{1}{x})(\text{Expected length fished out before return to the next fresh water interval of length } x \mid k^{th} \text{ order conflict})$$

The boundary conditions are $B_0 = B_1 = 1$. Where α denotes the fraction of the conflicted interval that one tries next in hopes of resolving the conflict (Gallager used $\alpha = 1/2$), the recursion is

$$B_k = P_{k,0}[\alpha+(1-\alpha)B_k] + P_{k,1}[\alpha+(1-\alpha)B_{k-1}] + \sum_{i=2}^{k} P_{k,i}\,\alpha B_i$$

or

$$(1-\bar{\alpha}P_{k,0}-\alpha P_{k,k})B_k = \alpha P_{k,0}+\alpha P_{k,1} + \bar{\alpha}P_{k,1}B_{k-1} + \alpha\sum_{i=2}^{k-1} P_{k,i}\,B_i \qquad (\bar{\alpha}=1-\alpha)$$

Then, the average length explored before two fish in a row are caught is

$$xB = x\sum_{k=0}^{\infty} q_k B_k$$

where again $q_k = x^k e^{-x/k!}$. The efficiency then is $\eta = xB/W$. Calculations reveal that the maximum value of η is 0.4877, which occurs for α slightly above 0.49. Thus, Gallager's algorithm ($\alpha = 1/2$) is only 0.0006 below optimum and therefore probably is preferable in practice because of its relative simplicity.

5. Upper Bounds

We present two arguments that lead to upper bounds for η_{max}. The first one, an information-theoretic argument devised by Pippenger,[2] shows that $\eta_{max} \leq 0.7448$. The second one, a genie argument attributable to Molle,[3] yields the stronger result $\eta_{max} \leq 0.6731$. It is felt that presentation of both results is merited because their derivations proceed along totally different lines. We conclude by indicating how we feel Molle's approach can be extended to yield $\eta_{max} \leq 0.5254$.

5.1. Pippenger's Upper Bound, $\eta_{max} \leq 0.7448$

Consider an algorithm for totally fishing out the interval [0,A], where A is a large number; eventually we shall let A approach infinity. Any such algorithm can be envisioned as embedded in a ternary decision tree. The three branches emanating from a node correspond to 0, 1 or ≥ 2 fish residing in the portion of the stream (not necessarily an interval) that we fish when at that node. Let p_k denote the probability that someone applying the algorithm in question will ever reach node k. (Although each p_k is a probability, note that $\{p_k\}$ is <u>not</u> a probability distribution.) Then the average number of trials required to fish out [0,A] with this algorithm is

$$W = \sum_k p_k \quad .$$

We need consider only algorithms for which $W < \infty$ for $A < \infty$.

Let q_{0k}, q_{1k}, and q_{2k} be the respective probabilities that someone applying the algorithm will find 0, 1, or ≥ 2 fish given that he is about to fish the portion of the stream associated with node k. Then the average number of fish caught in all is $\Sigma_k p_k q_{1k}$. Again assuming a stream of intensity 1, we have

$$\sum_k p_k q_{1k} = A$$

because all the fish in [0,A] eventually are caught and on the average there are A of them.

When one applies the algorithm, one follows a path through certain nodes of the decision tree. Finally, a terminal node, or "leaf", is reached, call it ℓ. At that juncture the entirety of [0,A] has been partitioned into a collection $C(\ell)$ of intervals each of which either is known not to have contained a fish or is known to have contained exactly one fish. Let the random variable L represent the index of the leaf at which the algorithm will terminate. Since each leaf is in one-to-one correspondence with the path through the tree from the root to said leaf, we can specify the value ℓ that L assumes by specifying the ternary sequence of successive branches the algorithm follows. The entropy of L therefore equals the average amount of information that must be supplied to specify which path was followed. Let H_k denote the average amount of information required to specify which branch is followed out of node k given that node k is visited.

By Fano's inequality[9] we know that

$$H_k \leq H(q_{1k}) + (1 - q_{1k})\log 2$$

where $H(q) = -q \log q - (1-q)\log(1-q)$ is the binary entropy function. It follows that the entropy of L is

$$H(L) = \sum_k p_k H_k \leq \sum_k p_k[H(q_{1k}) + (1-q_{1k})\log 2]$$

$$= \sum_k p_k H(q_{1k}) + (W - A)\log 2$$

If we let

$$p'_k = \frac{p_k}{\sum_k p_k} = \frac{p_k}{W} ,$$

then $\{p'_k\}$ is a probability distribution on the decision tree nodes. Since $H(q)$ is convex $\cap$, we deduce from Jensen's inequality that

$$H(L) \leq W \sum_k p'_k \, H(q_{1k}) + (W-A)\log 2$$

$$\leq WH(\sum_k p'_k q_{1k}) + (W-A)\log 2 ,$$

or

$$H(L) \leq WH(A/W) + (W-A)\log 2 .$$

It is also possible to bound $H(L)$ from below. Start by noting that

$$H(L) = \sum_{\ell} p(\ell)\log\frac{1}{p(\ell)} ,$$

where $p(\ell)$ is the probability that $L = \ell$ is the terminal node. Let $x_i = x_i(\ell)$, $1 \leq i \leq n(\ell)$, denote the lengths of the $n(\ell)$ subinternals of the partition $C(\ell)$ of $[0,A]$ associated with leaf ℓ that contain exactly one fish, and let $y_j = y_j(\ell)$, $1 \leq j \leq m(\ell)$, denote the lengths of those that contain no fish. Then $p(\ell)$ is less than or equal to the probability that the Poisson stream sample function has exactly one fish in each of said intervals of length x_i and none in each of those of length y_j, namely

$$p(\ell) \leq \prod_i x_i e^{-x_i} \prod_j e^{-y_j} = (\prod_i x_i) e^{-(\sum_i x_i + \sum_j y_j)}$$

$$p(\ell) \leq e^{-A} \prod_i x_i .$$

Thus, $H(L) \geq \sum_{\ell} p(\ell)\, [A \log e - \log \prod_{i=1}^{n(\ell)} x_i(\ell)]$

$$\geq A \log e - \sum_{\ell} p(\ell) \log \left[\frac{1}{n(\ell)} \sum_{i=1}^{n(\ell)} x_i(\ell)\right]^{n(\ell)}$$

where we have used the inequality between geometric and arithmetic means in the last step. Since $\Sigma_i x_i(\ell) \leq A$, we get

$$H(L) \geq A \log e + \sum_{\ell} p(\ell)\, n(\ell) \log(n(\ell)/A)$$

The function $f(x) = x \log(x/A)$ is convex $\cup$, so Jensen's inequality yields

$$H(L) \geq A \log e + \bar{n} \log(\bar{n}/A)$$

where $\bar{n} = \Sigma_{\ell}\, p(\ell)\, n(\ell)$. Since $n(\ell)$ is the total number of fish caught when the algorithm terminates at leaf ℓ, $\bar{n}$ is the average number caught when $[0,A]$ is fished out. Thus, $\bar{n} = A$ and we have

$$A \log e \leq H(L) \leq WH(A/W) + (W-A)\log 2$$

In the limit as $A \to \infty$, A/W becomes the "effishiency" of the algorithm in question. We deduce, therefore, that η for any algorithm must satisfy

$$\eta \log e \leq H(\eta) + (1 - \eta)\log 2.$$

Computation reveals that the largest value of η that satisfies this equation is 0.7448, thereby establishing the upper bound $\eta_{max} \leq 0.7448$.

5.2. Molle's Upper Bound, $\eta_{max} \leq 0.6731$

An entirely different approach to upper bounding η_{max} recently has been introduced by Molle.[3] Molle begins by approximating the Poisson stream by a Bernoulli stream. A Bernoulli stream is one that is divided into equal-length contiguous intervals each of which contains exactly one fish with probability p and is empty with probability 1-p. The Bernoulli stream becomes a Poisson stream of intensity λ in the limit as $M \to \infty$, $p \to 0$ and $Mp \to \lambda$ where M is the number of segments per unit length. Molle upperbounds η_{max} for the Bernoulli stream with parameter p via a "genie" argument and then studies the behavior of said bound in the Poisson limit. Since, unlike Molle, we have no inherent interest here in the Bernoulli approximations, we shall apply the genie argument to the Poisson limit directly rather than considering it as a limiting case.

Suppose that, in the event that we find the interval (0,x] to be congested, a beneficent genie pinpoints for us the exact locations a_1 and a_2 of the first two of the fish responsible for the conflict (cf. Figure 3).

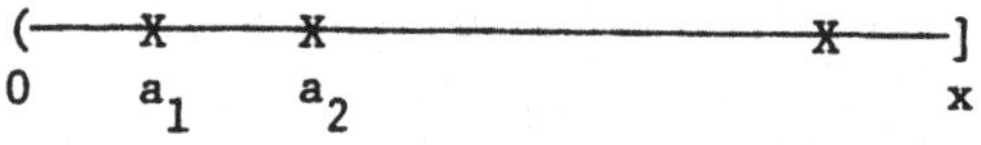

Figure 3. A congested basic interval with two genie-specified fish.

There may or may not be fish in [0,x] other than the ones at a_1 and a_2; the genie does not provide us with any information about this matter. It is not difficult to show that after the genie's information has been provided, a_2,x] reverts to fresh water. Therefore, after fishing $\{a_1\}$ and then $\{a_2\}$ to catch the two genie-specified fish, we explore a new fresh water interval of measure x, e.g., $(a_2, a_2 + x]$. So doing yields an empty net with probability e^{-x}, a fish with probability xe^{-x} and another conflict with probability $1 - (1+x)e^{-x}$. In the event of another conflict, the genie will help us again. If there is no conflict in $(0,x]$ in the first place, we go on to $(x,2x]$, and so on.

It is clear that the above procedure makes optimum use of the information provided by the genie. Therefore, the fraction of steps that yield a fish using this procedure serves as an upper bound to the optimum efficiency η_{max} attainable without the genie's help. This fraction is easily seen to be

$$\frac{xe^{-x} + 2[1-(1+x)e^{-x}]}{e^{-x} + xe^{-x} + 3[1-(1+x)e^{-x}]} = \frac{2-(2+x)e^{-x}}{3-2(1+x)e^{-x}}$$

Differentiation reveals that this is maximized at the solution of $3-x = 2e^{-x}$, namely $x \approx 2.887$, and that the maximum value is 0.6731. Therefore, $\eta_{max} \leq 0.6731$.

5.3. Conjectured upper bound, $\eta_{max} \overset{?}{\leq} 0.5254$

We now present a modification of the Poisson version of Molle's genie argument which appears to yield the considerably tighter result $\eta_{max} \leq 0.5254$; this is only 0.0377 above the lower bound of 0.4877 achieved by the optimized version of Gallager's algorithm discussed in Section 4.

Let us replace Molle's generous genie by one who gives less information. Then we will not be able to fish as efficiently as before and stand, therefore, to derive an improved (lowered) upper bound to η_{max}. In particular, in the event of a conflict in $(0,x]$, the new genie tells us a_2 of Fig. 3 but redistributes the two fish at a_1 and a_2 uniformly and independently over $G \triangleq (0,a_2]$. As was the case in Section 5.2, the Poisson statistics assure us that $(a_2,x]$ reverts to fresh water upon receipt of the genie's information. Hence, we need simply catch the 2 fish in G and then return to the basic fresh water problem.

Let W denote the least average number of steps required to catch the 2 fish in G. To see that W = 3 first recall that, if it is known that an interval in a Poisson process contains exactly n events, their occurrence times are distributed over the interval uniformly and mutually independently. Hence, W clearly is independent of the length of G. Let α denote the fraction of G explored at the first step. Then with probability α^2 we get a conflict and with probability $(1-\alpha)^2$ we come up empty. In either of these cases we are again faced with the problem of fishing out an interval known to contain exactly 2 fish; if we believed α was the right fraction to use, we should use it again. If we catch a fish, which happens with probability $2\alpha(1-\alpha)$, then we surely catch the other one next time and then return to the fresh water problem. Thus, we seek the value of α that minimizes $W(\alpha)$ where

$$W(\alpha) = 1+[\alpha^2+(1-\alpha)^2]\, W(\alpha) + 2\alpha(1-\alpha)$$

It is readily verified that $W(\alpha)$ is minimized for $\alpha = 1/2$, for which it equals 3.

It follows that, whereas the expected number of fish caught per cycle still is $xe^{-x} + 2[1-(1+x)e^{-x}]$, the expected number of steps it takes to catch them now is $e^{-x} + xe^{-x} + 4[1-(1+x)e^{-x}]$. The factor of 4 reflects the fact that each conflict now consumes an average of 4 steps - one to identify it and an average of 3 more to catch the two fish in the interval G specified by the genie. It follows that η_{max} is bounded from above by the maximum value of

$$\frac{xe^{-x} + 2[1-(1+x)e^{-x}]}{e^{-x} + xe^{-x} + 4[1-(1+x)e^{-x}]} = \frac{2-(2+x)e^{-x}}{4-3(1+x)e^{-x}}$$

Differentiation reveals that the maximum occurs at the solution of $4-2x = 3e^{-x}$, namely $x \simeq 1.735$, and that the maximum value is 0.5254.

The only reason why we advertise this result as a conjecture ($\eta_{max} \overset{?}{\leq} 0.5254$) instead of as a fact ($\eta_{max} \leq 0.5254$) is that it might be that one somehow can do better than simply to fish out G after the genie specifies it. Perhaps unions formed from portions of G and portions of fresh water should be considered. Intuition strongly suggests that this would be counter-productive, but a formal proof to that effect is lacking. In this connection we suspect that the following general "segregation conjecture" is valid not only for measurements that can distinguish only between 0, 1, and ≥ 2 fish but also for ones that can distinguish between 0, 1, 2, ..., n-1 and $\geq n$ fish, including the case $n=\infty$.

<u>Segregation Conjecture</u>. Let A and B be two disjoint sets in a Poisson stream. Assume that A and B are statistically independent given all the information I about the stream gathered so far in the sense that the

random vectors $(X_1, \ldots, X_m)$ and $(Y_1, \ldots, Y_n)$ are conditionally independent given I, where X_j denotes the number of fish in $A_j \subset A$ and Y_k denotes the number of fish in $B_k \subset B$. Further assume that neither A nor B is known to be devoid of fish and that at most one of them contains fresh water. Then the most efficient way to fish $A \cup B$ is to fish A and B separately.

The thesis by Mehravari[10] lends support to this conjecture by showing that a plausible "straddle algorithm" for fishing certain conditionally independent sets is slightly less efficient than fishing them separately. Proof of the segregation conjecture, which appears to be a challenging task, would among other things elevate the 0.5254 result from the status of an eminently reasonable conjecture to that of a true upper bound.

References

1. Gallager, R. G., Conflict resolution in random access broadcast networks, *Proc. of the AFOSR Workshop in Communications Theory and Applications*, Provincetown, MA, 17-20, 74, 17-20 September 1978.

2. Pippenger, Bounds on the performance of protocols for a multiple-access broadcast channel, Report RC-7742, Mathematical Sciences Department, IBM Thomas J. Watson Research Center, Yorktown Heights, NY, June 1979.

3. Molle, M. L., On the capacity of infinite population multiple access protocols, Computer Science Department, University of California, Los Angeles, CA, March 1980.

4. Roberts, L. G., Aloha packet system with and without slots and capture, ARPANET Satellite System Note 8, (NIC 11290), June 1972; reprinted in *Computer Communication Review*, Vol. 5, April 1975.

5. Mikhailov, V. A., On ergodicity of slotted ALOHA, Fifth International Symposium on Information Theory, Tbilisi, Georgia USSR, July 3-7, 1979.

6. Capetanakis, J. I., The multiple access broadcast channel: Protocol and capacity considerations, Technical Report ESL-R-806, MIT, Cambridge, MA, March 1978. (See also, Tree algorithm for Packet

broadcast channels, *IEEE Trans. on Information Theory*, Vol. IT-25, No. 5,. 505, September 1979.

7. Ruget, G., Lectures delivered at CISM summer school, July 1979, Udine, Italy.

8. Tsybakov, B. S., Berkovskii, M. A. Vvedenskaja, N. D., Mikhailov, V. A. and Fedorzov, S. P., Methods of random multiple access, Fifth International Symposium on Information Theory, Tbilisi, Georgia USSR, July 3-7, 1979.

9. Gallager, R. G., Information Theory and Reliable Communication, Wiley, New York, 1968.

10. Mehravari, N., The straddle algorithm for conflict resolution in multiple access channels, M.S. Thesis, Cornell University, School of Electrical Engineering, Ithaca, NY, May 1980.

PERFORMANCE ANALYSIS OF MULTI-USER COMMUNICATION SYSTEMS : ACCESS METHODS AND PROTOCOLS

Erol GELENBE
Université Paris-Sud
Laboratoire de Recherche en Informatique
91405 Orsay, France

1. Introduction

In a multi-user communication facility a set of processes (term which we use here in the sense of computer science, i.e. a "program" in execution) exchange information via a communication "channel". Various examples of such a system can be imagined : micro-processors exchanging data via a communication bus, transceivers using a common satellite channel, several programs exchanging messages through a packet switching network, etc. The rules imposed to the processes for their use of the communication medium have a fundamental influence on the usual performance measures (throughput, response time, reliability, etc.) of interest. However, the "second level" rules which the processes apply to their dialogue are also extremely important. These second level rules, usually called *protocols*, are designed to insure reliable communication of processes through an imperfect communication channel, and concern essentially the communicating processes, whereas the *access rules* define the relationship between the processes and the physical channel.

Higher level protocols also exist, which essentially concern larger ensembles of processes, or individual programs, making use of several levels of hardware and software in order to exchange information.

In these lectures we shall be primarily concerned with the first two levels : the access method to the channel and the protocols found at the second level. In the case of the access methods we shall essentially examine algorithms based on random access to a single channel : these methods were initiated in the ALOHA satellite packet switching network [1] and are now of great interest also for local area networks. The protocols we shall examine will cover two-process systems and we shall be concerned with simple but practical protocols such as send-and-wait and its generalizations and with HDLC.

2. Random access to a single broadcast channel

We consider a large set of terminals communicating over a common broadcast channel in such a way that a packet is successfully transmitted only if its transmission does not overlap in time with the transmission of another packet; otherwise all packets being simultaneously transmitted are lost. A packet whose transmission is unsuccessful is said to be blocked; it has to repeat the transmission until it succeeds. A packet which is not blocked is either active or transmitting a packet. The operation of the system is shown schematically in Figure 2.1, where the different state transitions of a packet are shown. Since the only means of communication between terminals is the channel itself, it is not easy to schedule transmissions so as to avoid collisions between packets.

Various methods for controlling the transmission of packets have been suggested. The simplest is to allow terminals to transmit packets at any instant of time. The second method, known as the slotted ALOHA scheme, has been shown to increase channel throughput over the first method [2]. Here, time is divided into "slots" of equal duration; each slot can accommodate the transmission of one packet, and packets are all of the same length. Pakcet transmission is synchronized so that it initiates at the beginning of a slot for any terminal and terminates at the end of the same slot. Another technique is based on the inhibition of transmissions (based on sensing the channel) when the channel is busy : such techniques are known as "carrier sense multiple access" (CSMA). In this chapter we will examine both the slotted ALOHA system, and the CSMA system.

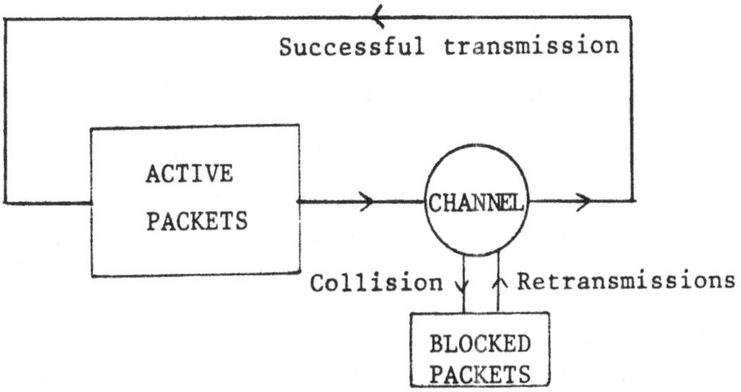

Figure 2.1 : Behaviour of packets in the broadcast channel.

2.1. Discrete time or slotted broadcast channels

In this section we shall examine the discrete time random access system. The following basic issues will be addressed :

(1) The uncontrolled behaviour of the system will be analyzed and its inherent instability will be established.

(2) Classes of policies leading to stable or "optimal" behaviour will be presented.

(3) Adaptive control policies which are practical approximations to the optimal control algorithms will be proposed and evaluated.

2.1.1. The stability problem

Kleinrock and Lam [3] have discussed the stability problem of the slotted ALOHA channel. Their results, based on simulations and fluid approximations, indicate that the channel becomes saturated if the set of terminals is very large, independently of the arrival rate of packets to the channel, saturation being the phenomenon whereby the number of blocked terminals becomes arbitrarily large. They also compute the expected time to attain a given level of saturation.

In this section we shall present a proof of instability based on our work [4,5] for an open (infinite set of terminals) or closed (finite set of terminals) system.

A precise definition of stability can be considered only in the context of a model of the behavior of the broadcast channel. Throughout this section we shall assume that a given terminal can transmit at most one packet at a time, thus we shall talk about a blocked or active terminal, rather than of a packet.

Assuming that the slot, and the time necessary to transmit a packet, are of unit length, we define $N(k)$ as the number of blocked terminals at the instant when the kth $(k=0,1,2,\ldots)$ slot begins. Let X_k be the number of packets transmitted from the set of active terminals during the kth slot, and denote by Y_k the number of blocked terminals transmitting during the kth slot. In the open model, (X_k) is a sequence of independent and identically distributed random variables with common probability distribution given by

$$\Pr(X_k = i) = c_i, \quad i \geq 0.$$

In the closed model, we let M denote the number of terminals in the system, and we assume that the event $(X_k = i \mid N(k) = j)$ is independent of values of X_t for $t < k$; its probability is given by

$$q_j(n) = \Pr(X_k = j \mid N(k) = n) = \binom{M-n}{j} b^j (1-b)^{M-n-j}$$

for $0 \leq j \leq M-n$, where b is the probability that any one active terminal transmits a packet during a slot.

For both models, we denote by f the probability that any one blocked terminal transmits a packet during a slot. We then define

$$(2.1) \qquad g_i(n) = \Pr(Y_k = i \mid N(k) = n),$$

where we assume that the event $(Y_k \mid N(k))$ is independent of Y_t for $t < k$. Therefore

$$(2.2) \qquad g_i(n) = \binom{n}{i} f^i (1-f)^{n-1},$$

and more particularly

$$(2.3) \qquad g_o(n) = (1-f)^n, \quad g_1(n) = nf(1-f)^{n-1}.$$

<u>Definition.</u> The infinite source broadcast channel is unstable if for $k \to \infty$ the probability $\Pr(N(k) < j) \to 0$ for all finite values of j; otherwise it is stable. For the finite source model, the system is unstable if the above condition holds as we let $M \to \infty$, $b \to 0$, $M \cdot b \to d$, where d is a constant.

The definition given here simply states that instability exists if (with probability one) the number of blocked terminals becomes infinite as time tends to infinity.

<u>THEOREM 2.1.</u> The broadcast channel is unstable for both the finite and infinite source models.

<u>PROOF.</u> Let us first consider the infinite source model. Let $p_n(k)$ denote the probability that $N(k) = n$.

The following balance equation may be written for the infinite source model :

$$(2.4) \qquad p_n(k+1) = \sum_{j=2}^{n} p_{n-j}(k) c_j + p_{n+1}(k) g_1(n+1) c_o + p_n(k)(1-g_1(n)) c_o$$
$$+ p_n(k) g_o(n) c_1 + p_{n-1}(k)(1-g_o(n-1)) c_1 .$$

To illustrate the interpretation of the right-hand side, we note that the first term covers the cases where two or more packets have been transmitted by the active terminals during the kth slot and the second term covers the case in which exactly one blocked terminal has transmitted while no active terminal has done so. Notice that $\{N(k);\ k=0,1,\dots\}$ is a Markov chain and that it is aperiodic and irreducible. It is ergodic if an invariant probability measure $\{p_n : n=0,1,\dots\}$ exists such that $p_n > 0$ for all n and $p_n = \lim_{k\to\infty} p_n(k)$. To show that $\lim_{k\to\infty} \Pr(N(k) < j) = 0$ for all finite values of j, it suffices to show that the Markov chain representing the number of blocked terminals is not ergodic. Setting $p_n = \lim_{k\to\infty} p_n(k)$ if it exists we obtain

$$(2.5)\qquad p_n = \sum_{j=0}^{n} p_{n-j}c_j + p_{n+1}g_1(n+1)c_o + p_n(g_o(n)c_1 - g_1(n)c_o) - p_{n-1}g_o(n-1)c_1.$$

Letting

$$(2.6)\qquad S_N = \sum_{n=0}^{N} p_n,$$

we then have, for any $N \geq 0$,

$$(2.7)\qquad S_N = p_{N+1}g_1(N+1)c_o + p_N g_o(N)c_1 + \sum_{n=0}^{N} S_{N-n}c_n$$

or

$$S_N(1-c_o) = \sum_{n=1}^{N} S_{N-n}c_n + p_{N+1}g_1(N+1)c_o + p_N g_o(N)c_1$$

or equivalently

$$(2.8)\qquad p_N(1-c_o) \leq p_{N+1}g_1(N+1)c_o + p_N g_o(N)c_1.$$

But then, from (2.3) and (2.8), we have

$$p_{N+1}/p_N \geq (1-c_o-(1-f)^N c_1)/(N+1)f(1-f)^N c_o$$

for any nonnegative integer N. This result implies that the ratio

$(p_{N+1}/p_N) \to \infty$ as $N \to \infty$; so the sum S_∞ can exist only if $p_N=0$ for all finite values of N - otherwise S_∞ is divergent, which cannot be the case when the p_N, $N \geq 0$, define a probability distribution. Thus the Markov chain representing the number of blocked terminals is not ergodic, and the broadcast channel under the infinite source assumption is unstable.

Now consider the finite source model. The balance equation for $0 \leq n < M$ is :

$$p_n(k+1) = \sum_{j=2}^{n} p_{n-j}(k)q_j(n-j) + p_{n+1}(k)g_1(n+1)q_o(n+1) + p_n(k)(1-g_1(n))q_o(n) + p_n(k)g_o(n)q_1(n) + p_{n-1}(k)(1-g_o(n-1))q_o(n-1). \tag{2.9}$$

Defining, for $0 \leq N < M$, the sum S_N as in (2.6) for the finite source model, we write

$$S_N = p_{N+1}g_1(N+1)q_o(N+1) + p_Ng_o(N)q_1(N) + \sum_{n=0}^{N} \sum_{j=0}^{n} p_{n-j}q_j(n-j). \tag{2.10}$$

Notice first that since the Markov chain is aperiodic, irreducible, and finite, it is ergodic for each $M < \infty$. Therefore $p_i > 0$, $0 \leq i \leq M$, $\Sigma_{i=1}^{M} p_i = 1$. With $N = M-1$, (2.10) yields

$$S_{M-2} - \sum_{n=0}^{M-2} \sum_{j=0}^{n} p_{n-j}q_j(n-j) + bp_{M-1}[1-(1-f)^{M-1}] = p_M f(1-f)^{M-1},$$

or, as can be easily verified,

$$(p_M/p_{M-1}) \geq b[1-(1-f)^{M-1}]/f(1-f)^{M-1}.$$

Therefore as $M \to \infty$ this ratio tends to infinity, which can only imply that $p_{M-1} \to 0$. But the argument is valid for any p_N/p_{N-1}, $0 < N \leq M$. Therefore as $M \to \infty$ we have $p_N \to 0$, $0 < N < M$, and $p_M \to 1$. This result proves the theorem also for the case of the finite source . □

The instability of the system has an immediate consequence on its throughput, or number of packets per unit time which can be transmitted successfully.

Definition. The conditional throughput $D_n(k)$ of the broadcast channel is the conditional probability that one packet is successfully transmitted during the kth slot given that $N(k) = n$.

Definition. The throughput of the broadcast channel is defined as

$$D = \lim_{k\to\infty} \sum_{n=0}^{\infty} D_n(k)\; p_n(k).$$

The conditional throughput is $D_n(k) = c_0 g_1(n) + c_1 g_0(n)$ for the infinite source model; for the finite source model we replace c_0 and c_1 by $q_0(n)$ and $q_1(n)$, respectively. This quantity is obviously independent of k; therefore in the following we simply write D_n instead of $D_n(k)$.

THEOREM 2.2. For fixed $f > 0$, the throughput of the broadcast channel is 0 for the infinite source model, and for the finite source model, as we let $M \to \infty$, $b \to 0$, $M\cdot b \to d$.

The proof is a straightforward consequence of Theorem 2.1.

On Figure 2.2 we show simulation results concerning the behaviour of the infinite source unstable channel : the number of blocked packets (N(k) and the instantaneous throughput D(k) (number of successful transmissions per unit time) are plotted as a function of k. The instability of the system exhibits itself as follows : beyond a certain (random) value of k, the number of blocked packets increases rapidly in the manner of an avalanche, since as this number increases the blocked packets' retransmissions provoque the blocking of even more packets. Simultaneously with this effect, the useful throughput of the system decreases drastically becoming, for all practical purposes, close to zero.

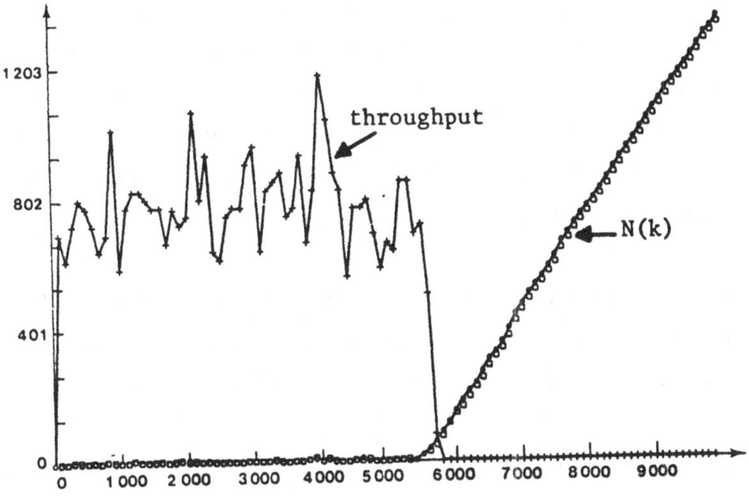

Figure 2.2. Simulation of the throughput and of the number of blocked terminals in the unstable channel.

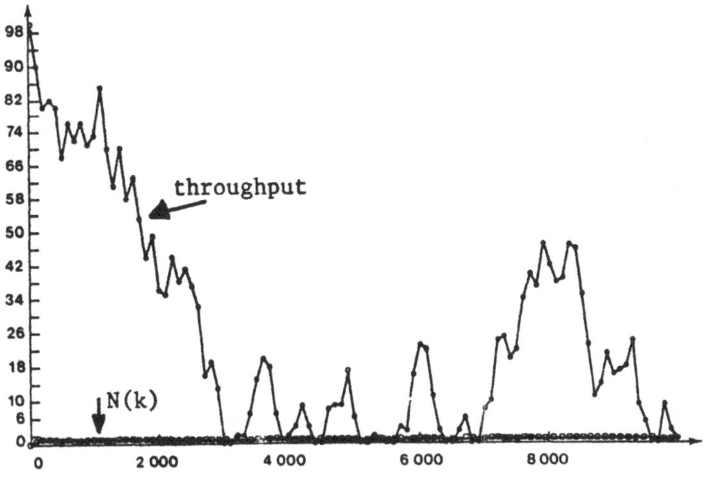

Figure 2.3. Simulation of the stable channel controlled by the optimal retransmission control policy.

2.1.2. A state dependent control policy

Obviously the performance of the system operating with a constant retransmission probability f is unacceptable, control policies must be designed which modify f as a function of the system state variables so as to obtain satisfactory performance. One such class of policies based on letting f depend on the number of blocked packets n will be examined in this section. These results were first published in [5].

We shall borrow the terminology of the control theory of Markov chains : a control policy is stationary if its decisions at a given instant of time are only based on the value of system state at that time.

THEOREM 2.3. A stationary control policy based on modifying f, i.e. a control policy which chooses a value f(k) during the kth slot based only on N(k), yields a stable channel if $\lambda < d$ and an unstable one if $\lambda > d$, where

$$\lambda \equiv \Sigma_1^\infty i c_i, \quad d \equiv \lim_{n\to\infty} [c_1 g_0(n) + c_0 g_1(n)],$$

for an open system, assuming that the limit expressed in d exists.

This theorem yields precise conditions for stability. In fact Theorem 2.1 is a special case, as we shall presently see, of Theorem 2.3. We shall only provide a proof of stability if $\lambda < d$; for the rest of the reader is referred to [4].

PROOF OF STABILITY

In order to prove that the system is stable if $\lambda < d$ we shall call upon the following lemma (due to Pakes [6]). Let $\{X_k\}_{k \geq 1}$ be a discrete time Markov chain, whose state-space is the set of non-negative integers, which is irreducible and aperiodic. It is ergodic, i.e., a positive stationary probability distribution $\{p_n\}$ exists such that

$$p_n = \lim_{k\to\infty} P[X_k=n] > 0, \; n \geq 0$$

if

(i) $E[X_{k+1} - X_k \mid X_k=n] < \infty$ for all $n \geq 0$,

(ii) $\limsup_{n \to \infty} E[X_{k+1} - X_k \mid X_k=n] < 0$

Turning now to the Markov chain $\{N(k)\}_{k \geq 1}$, it is clearly irreducible and aperiodic, thus we have have to verify conditions (i) and (ii). Notice that

$$N(k+1) = \begin{cases} N(k) + X(k), & \text{if } X(k) \geq 2 \text{ or} \\ & \text{if } X(k) = 1 \text{ and } Y(k) \geq 1 \\ N(k), \text{ if } X(k) = 1 \text{ and } Y(k) = 0 & \\ \quad \text{or if } X(k) = \sigma \text{ and } Y(k) \neq 1 & \\ N(k) - 1 \quad \text{if } X(k) = 0 \text{ and } Y(k) = 1 & \end{cases}$$

Condition (i) is obviously satisfied. Thus we have to prove that (ii) is equivalent to the condition $\lambda < d$ (if d, expressed as a limit, exists). Using the above equation we have that

$$E[N(k+1) - N(k) \mid N(k) = n]$$

$$= \Sigma_2^\infty jc_j + c_1(1-g_o(n)) - c_o g_1(n)$$

$$= \lambda - [c_1 g_o(n) + c_1 g_o(n)]$$

If the limit $d \equiv \lim_{n\to\infty} [c_1 g_o(n) + c_1 g_o(n)]$ exists, it means the lim sup and the lim inf of the expression are identical, so that condition (ii) becomes

$$\lambda < d$$

which had to be proved.

Corollary. Suppose that $0 < f < 1$ is a constant. Then we can easily see that $d=0$ leading to the result of Theorem 1.

2.1.3. An optimal policy $f^*(n)$

If an appropriate policy $f(n)$ leads to a stable system for all λ with $\lambda < d$, it is of interest to choose $f(n)$ so that d will be maximum. This "optimality" is understood as the choice of an $f^*(n)$ which maximizes the size of the set of λ's for which the system remains stable. Obviously, for a stable system the "output rate" or rate of successful packet transmissions will simply be the input rate λ. We shall proceed as follows in order to compute the "optimal" f^*. Call

$$d(n) = c_1 g_0(n) + c_0 g_1(n)$$

For each value of n we shall seek the value of f, call it $f^*(n)$, which maximizes $d(n)$. Taking the partial derivative of $d(n)$ with respect to f and setting it to zero, we obtain

$$f^*(n) = \frac{c_0 - c_1}{nc_0 - c_1} \tag{2.11}$$

and can verify that

$$\left.\frac{\partial^2 d(n)}{\partial f^2}\right|_{f=f^*(n)} < 0$$

It is of interest to compute the optimum value of $d(n)$ and of d. Using (2.11) we obtain

$$d^*(n) = c_0 \left[\frac{n-1}{n-c_1/c_0}\right]^{n-1} \tag{2.12}$$

and

$$d^* = \exp\left(\log c_0 + \frac{c_1}{c_0} - 1\right) \tag{2.13}$$

If the arrival process of fresh packets is Poisson we have

$$d^*(n) = e^{-\lambda} \left[\frac{n-1}{n-\lambda}\right]^{n-1} , \quad d^* = 1/e$$

2.2. A "threshold" control policy

Another state dependent control policy suggested in [4] is based on limiting the access of fresh packets to the network in addition to scheduling retransmissions. This policy will be called a threshold control since its principle is to inhibit access if the number of blocked packets has attained a predetermined level n_o.

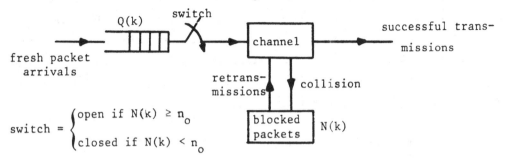

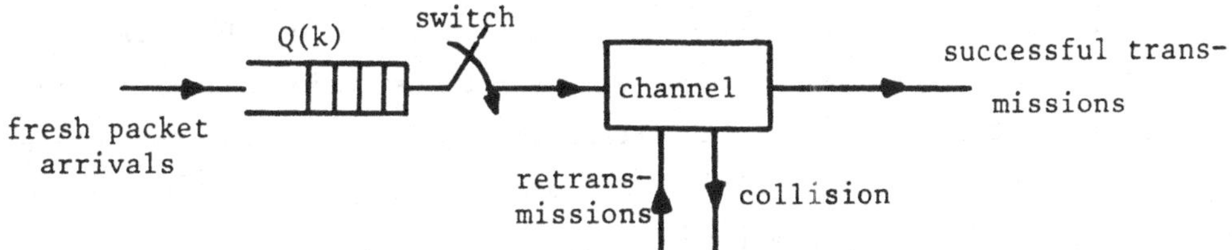

$$\text{switch} = \begin{cases} \text{open if } N(k) \geq n_o \\ \text{closed if } N(k) < n_o \end{cases}$$

Figure 2.4 : Schematic representation of the threshold control policy.

We shall analyze this control policy with respect to its stability and its optimality as in the case of the retransmission control policy of the previous sections. However the principal measure of interest cannot anymore be the number of blocked terminals N(k), since Q(k) packets also wait in the input queue to the channel. We shall therefore examine the process L(k) = N(k) + Q(k). The definition we previously gave about stability will be now extended to $\{L(k)\}_{k \geq 1}$.

Blocked terminals will retransmit independently of each other with probability f in each slot as with the retransmission control policy.

THEOREME 2.4. The threshold control policy is stable for the open model if $\lambda < \tilde{A}$ and unstable if $\lambda > \tilde{A}$, where $\tilde{A} = g_1(n_o)$.

Again, we shall only prove the result for $\lambda < A$ using Pakes' lemma. The proof of instability if $\lambda > A$ can be found in [4].

PROOF OF STABILITY From the description of the threshold control policy, we can write for $L(k) > n_o$:

$$L(k+1) = L(k) + X(k) - 1[Y(k) = 1] \tag{2.14}$$

where $1[\mathcal{a}]$ takes the value 1 if $\mathcal{a}$ is true, and is zero if $\mathcal{a}$ is not true. Clearly, if $L(k) > n_o$, then $N(k) = n_o$ and $Q(k) = L(k) - n_o$. If $L(k) \leq n_o$ the equations are identical to those of the retransmission control policy. It is clear that $\{L(k)\}_{k \geq 1}$ is an irreducible and aperiodic Markov chain. It is also clear that (for finite λ, of course) $E[L(k+1)-L(k)|L(k)=n] < \infty$ for all finite n. Furthermore from (2.14)

$$\lim_{n\to\infty} E[L(k+1)-L(k)|L(k)=n] = \lambda - g_1(n_o)$$

which proves the stability condition $\lambda < g_1(n_o)$.

A similar "optimality" problem arises here. The problem is now to choose f, the retransmission probability for blocked terminals, in order to maximize $g_1(n_o)$. We shall first compute the value f^*_{th} of f which maximizes $g_1(n_o)$: we can easily see that it is

$$f^*_{th} = 1/n_o. \tag{2.15}$$

We see that f^*_{th} has a form similar to that of the retransmission control policy studied earlier, except that now it has a fixed value independent of the number of blocked terminals. For large values of n_o we shall have

(2.16) $$g_1(n_o)\Big|_{f=f^*_{th}} \cong 1 - \frac{1}{n_o}$$

The threshold control policy thus has the desirable property that if we are able to implement it properly it can provide a useful throughput close to unity.

2.4. Adaptive implementation of the optimal retransmission control policy

In this section we shall examine an adaptive control scheme which allows us to implement approximately the optimal control algorithm for retransmissions based on setting f close to its optimal value

$$f^* = (c_o-c_1)/(nc_o-c_1)$$

The main difficulty with this scheme is that the parameter n, which is the value of the total number of blocked terminals, is unavailable to the set of blocked terminals who have themselves use of this information. Our algorithm will therefore have to be constructed from the only information available to all of the terminals :

- the idle slots when no transmission has occured,
- the slots in which exactly one (successful) transmission takes place.

When a collision takes place it may not be possible to count the total traffic, i.e. the total number of packets which have taken part in the collision cannot be known precisely.

Denoting as usual by N(k) the number of blocked terminals in the system at the beginning of the kth slot, by Y(k) the number of retransmissions from them and X(k) the number of transmissions from the active terminals in slot k, we can built a sequence of r.v. (random variables) $\{T(k)\}_{k \in N}$ with the r.v. X(k) and Y(k) :

$$T(k) = X(k) + Y(k) , \qquad k \geq 1$$

$T(k)$ is the total traffic in the channel during slot k. The conditional traffic T_n is defined as

$$T_n = E\{T(k)|N(k)=n\}$$

THEOREM 2.5. For the broadcast channel under optimal retransmission policy f (n), we have :

$$T_n = 1 + \lambda - c_1/c_o + (c_1/nc_o)(1-c_1/c_o) + O(1/n^2)$$

If we add the assumption that the active terminals form a Poisson source, we obtain :

$$T_n = 1 + (\lambda/n)(1-\lambda) + O(1/n^2)$$

This result can easily be derived from the definitions of T_n, c_i.

The expressions of T_n given above are very interesting, since they tell us that T_n tends to a constant T^* independent of n as soon as the number n of blocked terminals is high enough (practically when n > 5). If the active terminals form a Poisson source then $T^* \cong 1$. The following algorithm was introduced in [8].

We wish to estimate as accurately as possible $T(k)$. We would then like to set the retransmission probabilities so that $T(k)$ approaches T^*.

By "listening" to the channel, we can distinguish silent slots from the others and define the variable $S(k)$:

$$S(k)=\begin{cases} 1 & \text{if } T(k) = 0 \\ 0 & \text{otherwise} \end{cases} \qquad k \geq 1$$

Dividing the channel time into "windows" of L slots, we then define

$$S'(k) = \sum_{j=0}^{L-1} S(k-j)/L \qquad 0 < L < \infty,\ L-1 < k < \infty$$

$$\hat{T}(k) = -\text{Ln } S'(k)$$

If the process $\{T(k)\}_{k \geq 1}$ is Poisson and stationary then $S'(k)$ is the maximum likelihood estimator of $\Pr[T(k)=0]$ and $\hat{T}(k)$ is an estimator of $T(k)$.

Let $\{f_k\}_{k \geq 1}$ be a sequence of random variables taking values in $]0,1[$ and which will give the values of the retransmission probability in the adaptive control. The process $\{f_k\}_{k \geq 1}$ is chosen so as to depend on th the process $\{\hat{T}(k)\}_{k \geq 1}$ in the following way :

$$f_{k+1} = \begin{cases} f_k \dfrac{\ell}{\ell-1}, & \text{if } \hat{T}(k) \leq 1-\varepsilon_1 \\ f_k, & \text{if } 1-\varepsilon_1 < \hat{T}(k) < 1+\varepsilon_2 \\ f_k \dfrac{\ell}{\ell+1}, & \text{if } \hat{T}(k) \geq 1+\varepsilon_2 \end{cases}$$

where $\varepsilon_1, \varepsilon_2 > 0$ are parameters whose values determine the speed of reaction of the algorithm and ℓ is an integer characterizing the control.

The simple form $f^*(n) = (c_o-c_1)/(nc_o-c_1)$ of the optimal control is being now replaced by a more complicated one which depends on the set of parameters $\{\varepsilon_1, \varepsilon_2, \hat{T}(k), \ell\}$.

Instead of computing $S'(k)$ as a moving average, we may also consider, with a view to saving processing time, the possibility of jumping from window to window and keeping the same control throughout a given window. A refinement can also be considered, which consists in exponentially smoothing the computation of $S'(k)$: we can then decide at what speed the influence of the past history should fade. These two modifications have not been examined in these lectures but are reported on elsewhere [7].

This adaptive algorithm has been simulated : some results are shown in Figures 2.5 and 2.6 where the actual number of blocked terminals N(k) is plotted as a function of time. From the value f_k of the retransmission probability, computed by the estimator, we have calculated the "estimated" value $\hat{N}(k)$ of the number of blocked terminals :

$$\hat{N}(k) = [\hat{T}(k)-\lambda]/f_k$$

We can see the excellent agreement between $\hat{N}(k)$ and N(k), the measured value.

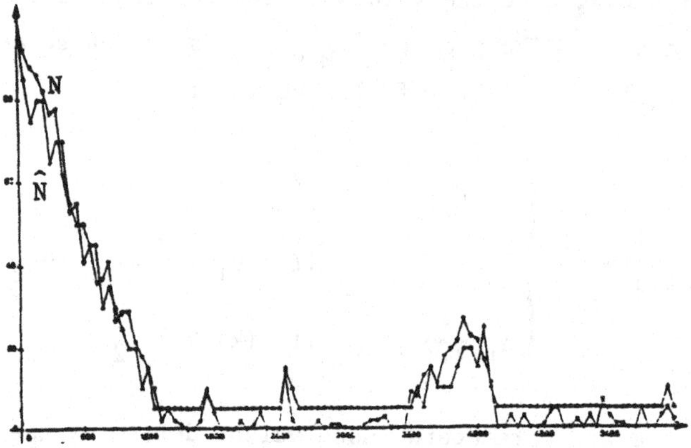

Figure 2.5. Adaptive control with λ=0.3; $(\varepsilon_1,\varepsilon_2)$=(0.20,0.80)

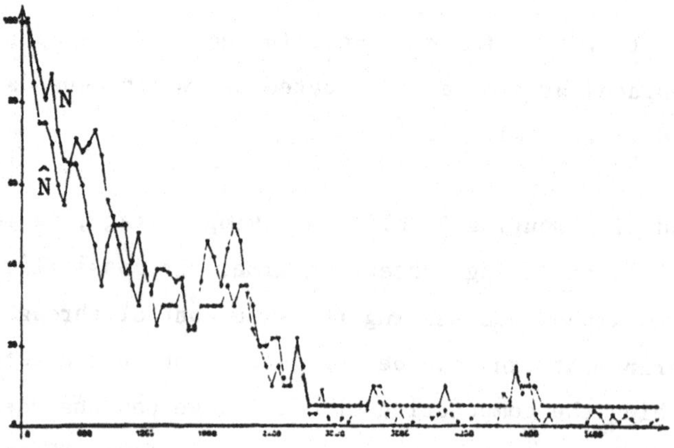

Figure 2.6. Adaptive Control with λ=0.3; $(\varepsilon_1,\varepsilon_2)$=(0.20,1.20)

REFERENCES

1. ABRAMSON N., "The ALOHA system - another alternative for computer communications", Proc. AFIPS 1970 FJCC, Vol. 37, AFIPS Press., Montvale, N.J., pp. 281-285.

2. ABRAMSON N., "Packet switching with satellites", Proc. AFIPS, 1973, NCC, AFIPS Press., Montvale, N.J., pp. 695-702.

3. KLEINROCK L., LAM S., "Packet switching in a slotted satellite channel", Proc. AFIPS 1973 NCC, AFIPS Press., Montvale, N.J., pp. 703-710.

4. FAYOLLE G., GELENBE E., LABETOULLE J., "Stability and optimal control of the packet swithcing broadcast channel", Journal ACM, 24, 3, pp. 375-386, 1977.

5. FAYOLLE G., GELENBE E., LABETOULLE J., "The stability problem of broadcast packet switching computer networks", ACTA INFORMATICA 4, 1974, pp. 49-53.

6. PAKES A.G., "Some conditions for ergodicity and recurrence of Markov chains", Oper. Res. 17, 1969.

7. BANH T.A., "Réseaux d'ordinateurs à commutation de paquets : modélisation mathématique et simulation en vue de l'optimisation des performances", Thèse de Doctorat en Sciences Appliquées, 1977-1978, Liège University, Belgium.

9. KLEINROCK L., LAM S., "Packet switching in a multi-access broadcast channel : Performance evaluation", IEEE Trans. on Communications COM-23, 4 (April 1975), pp; 410-423.

10. LAM S., Ph.D.Th., Comptr., Sci., Dep., U. of California, Los Angeles, 1974.

11. METCALFE R.M., "Steady-state analysis of a slotted and controlled ALOHA system with blocking", Proc. Sixth HawaïInt. Conf. on Syst. Sci. Honolulu, Jan. 1973, pp. 375-378.

12. METCALFE R.M., and BOGGS D.R., "ETHERNET : Distributed packet switching for local computer networks", CACM, July 1976, Vol. 19, N°7, pp. 395-404.

13. SPANIOL O., "Mehrfrequenz-ALOHA-Netzwerke", Informatik-Fachberichte N° 9, Springer-Verlag, pp. 277-295, Proc. of the Conf. "Workshop über Modelle für Rechensysteme", 31/3-1/4/1977. Bonn, RFA.

14. TOBAGI F.A., KLEINROCK L., "Packet switching in Radio channels : Part III - Polling and (dynamic) split-channel reservation Multiple Access", IEEE Trans. on Communications, Vol. COM-24, N°8, pp. 832-845, August 1976.

ABOUT A COMBINATORIAL PROOF OF THE NOISY CHANNEL CODING THEOREM

János Körner

Mathematical Institute of the
Hungarian Academy of Sciences

Introduction

The most famous problem of information theory, that of determining the zero-error capacity of a discrete memoryless channel, is of combinatorial nature. Originally stated by Shannon [1] in 1956, it has been studied by many combinatorialists. In a recent paper Lovász [2] developed a sophisticated method to derive converse results on the zero-error capacity and succeeded to settle an intriguing special case. This is a channel of which the five input letters can be arranged cyclically so that two input letters can result in a same output letter with positive probability iff they are adjacent in this cyclical array. This "pentagon" constitutes the simplest case for which Shannon was unable to determine the zero-error capacity in 1956. Unfortunately, the Lovász bounding technique also fails in many important cases, cf. Haemers [3]. It has often been argued that the problem is not intrinsically information-theoretic, since it can be stated without using probabilistic concepts. (This argument was even brought up as an excuse for the information theorists' inability to solve the problem.) In the last couple of years, however, an increasing number of people seem to believe that in the discrete case, all the classical results of information theory can be rederived using combinatorial methods. Moreover, the proofs so obtained often are simpler and more intuitive than earlier ones. The present tutorial paper should propagate this belief.

We shall consider the classical problem of determining the capacity of a discrete memoryless channel. Using a very simple and entirely combinatorial version of the method of typical sequences we shall derive surprisingly sharp exponential error bounds on the probability of error. The coding theorem and its strong converse will be obtained as simple corollaries. Then, we shall comment on the universal coding aspects of what we have proved.

Neither the results nor the method are new. As everybody knows, the Noisy Channel Coding Theorem was proved by Shannon [4], [5], and the strong converse was obtained by Wolfowitz [6], cf. also [7]. The present proofs featuring exponentially decreasing error terms are due to Dueck-Körner [8] and Csiszár-Körner [9]. For the history of exponential error bounds, the reader is advised to consult [8] and [9]. At present, we just mention that the random coding bound we shall obtain is originally due to Fano [10] whose proof was simplified by Gallager [11]. The proof in [9] is in the line of Csiszár-Körner-Marton [12]. The strong converse with an exponential error term is due to Arimoto [13], who obtained the exponent in a different algebraic form. The present form of the exponent appears in [8].

Statement of the Problem

A *discrete memoryless channel* (DMC) is given by a stochastic matrix $W: \mathcal{X} \to \mathcal{Y}$. Here $\mathcal{X}$ and $\mathcal{Y}$ are arbitrary finite sets, called the *input alphabet* and the *output alphabet,* respectively. For every positive integer n, the matrix W has an *n'th memoryless extension* $W^n: \mathcal{X}^n \to \mathcal{Y}^n$. Here $\mathcal{X}^n$ is the set of sequences of length n which can be formed from the elements of $\mathcal{X}$. The entries of the matrix W^n are obtained from those of W by the formula

$$W^n(\mathbf{y}|\mathbf{x}) \triangleq \prod_{i=1}^{n} W(y_i|x_i)$$

in which $\mathbf{y} = y_1 y_2 \ldots y_n$ and $\mathbf{x} = x_1 x_2 \ldots x_n$.

A *block code* of block-length n for discrete memoryless channels with input alphabet $\mathcal{X}$ and output alphabet $\mathcal{Y}$ is a pair of mappings (f, φ) such that f maps some finite set $\mathcal{M}$ into $\mathcal{X}^n$, while φ maps $\mathcal{Y}^n$ back into some set $\mathcal{M}'$ containing $\mathcal{M}$. The elements of $\mathcal{M}$ are called *messages,* f is called an *encoder* and φ is a *decoder.* The images f(m), $m \in \mathcal{M}$ of the messages under f are the *codewords.* This terminology refers to communication, although our problem will be entirely mathematical. For a given message $m \in \mathcal{M}$ we shall say that a *transmission error* occurs, if the transmission of the codeword f(m) over the DMC results in a sequence $\mathbf{y}$ such that $\varphi(\mathbf{y}) \neq m$. The *probability of erroneous transmission of message m* is

$$e(m) = e(W^n, f, \varphi, m) \triangleq 1 - W^n(\varphi^{-1}(m)|f(m)), \qquad (1)$$

where

$$W^n(\varphi^{-1}(m)|f(m)) = \sum_{y:\varphi(y)=m} W^n(y|f(m))$$

is the total conditional probability of the event that the transmission of codeword f(m) will result in a sequence **y** which is decoded into message m. The *average probability of error* $\bar{e}$ of the code (f, φ) is defined as the arithmetic average of e(m) as m varies in the message set, i.e.

$$\bar{e} = \bar{e}(W^n, f, \varphi) = \frac{1}{|\mathcal{M}|} \sum_{m \in \mathcal{M}} e(W^n, f, \varphi, m) . \tag{2}$$

It is intuitively clear that for a fixed block length n, the smaller $\bar{e}$, the smaller the cardinality of $\mathcal{M}$. The coding problem consists exactly in establishing the payoff of these two basic performance criteria of a code. To make this precise, let us call the quantity

$$R_f \triangleq \frac{1}{n} \log |\mathcal{M}|$$

the *rate of the code* (f, φ). There are two equivalent ways to look at this payoff. Fixing a threshold $\epsilon > 0$, one can ask for the largest possible rate of a block code having average probability of error at most ϵ. More precisely, one says that R is an *ϵ-achievable rate* if for every $\delta > 0$ and large enough n there exist block codes of block length n such that

$$R_f \geq R + \delta, \quad e(W^n, f, \varphi) \leq \varepsilon . \tag{3}$$

The largest ϵ-achievable rate C_ϵ is called the *ϵ-capacity* of the DMC. The number

$$C \triangleq \inf_{\varepsilon > 0} C_\varepsilon$$

is called the *capacity* of the DMC. (At this point we are able to formulate Shannon's zero-error capacity problem. As the reader can guess, this consists in determining C_0.) This problem statement is, however, not entirely satisfactory. Namely, it is not clear why one should keep the threshold ϵ constant as n tends to infinity. Instead, one could fix an arbitrary sequence $\epsilon_n \to 0$ and ask for the determination of the largest R such that there is a sequence of block codes (f_n, φ_n) with $R_{f_n} \to R$ and $\bar{e}(W^n, f_n, \varphi_n) - \epsilon_n \to 0$. For reasons of tradition, those interested in this more refined question have always preferred to ask it from the reversed direction. We shall say that $E \geq 0$ is an *attainable error exponent at rate R* for the DMC given by $W: \mathcal{X} \to \mathcal{Y}$ if for every $\delta > 0$ and sufficiently large n there exist block codes (f, φ) of block length n satisfying

$$(4) \qquad R_f \geq R - \delta , \qquad \frac{1}{n} \log \bar{e}(W^n, f, \varphi) \leq -E + \delta .$$

For every R, denote by E(R,W) the largest E that is an attainable error exponent at rate R for the given DMC. E(R,W) as a function of R is called the *reliability function* of the DMC. We know that by the definition of capacity,

$$(5) \qquad R > C \quad \text{implies} \quad E(R, W) = 0 .$$

Thus, for rates R above capacity another question is in order. Let us say that $G \geq 0$ is an *attainable exponent of correct decoding at rate* R for the DMC given by $W: \mathcal{X} \to \mathcal{Y}$ if for every $\delta > 0$ and sufficiently large n there exist block codes (f, φ) of block length n yielding

$$(6) \qquad R_f \geq R - \delta, \qquad \frac{1}{n} \log(1 - \bar{e}(W^n, f, \varphi)) \geq -G - \delta .$$

Similarly to the foregoing, for every R we denote by G(R,W) the smallest G that is an attainable exponent of correct decoding at rate R for the given DMC. Considered as a function of R, G(R,W) will be called the *reliability function of the DMC at rates above capacity.* By the definition of capacity,

$$(7) \qquad R < C \quad \text{implies} \quad G(R, W) = 0 .$$

We shall derive lower bounds both on E(R,W) and G(R,W), proving also that except for trivial cases

$$(8) \qquad R < C \quad \text{implies} \quad E(R, W) > 0$$

$$(9) \qquad R > C \quad \text{implies} \quad G(R, W) > 0 .$$

This will establish the coding theorem and its strong converse in a very sharp formulation. It is worthwhile to add that a precise formula for E(R,W) would imply the determination of the zero-error capacity C_0.

Preliminaries

To make this paper entirely self-contained, we recall some of the basic definitions of information theory.

The *entropy* of a probability distribution P on $\mathcal{X}$ is denoted as

$$H(P) \triangleq \sum_{x \in \mathcal{X}} P(x) \log \frac{1}{P(x)} .$$

For a stochastic matrix $V: \mathcal{X} \to \mathcal{Y}$ we denote by $H(V|P)$ the average entropy of the rows of V averaged by the distribution P on $\mathcal{X}$, i.e.

$$H(V|P) \triangleq \sum_{x \in \mathcal{X}} P(x)\, H(V(\cdot|x)) = \sum_{x \in \mathcal{X}} P(x) \sum_{y \in \mathcal{Y}} V(y|x) \log \frac{1}{V(y|x)} .$$

More intuitively $H(V \mid P)$ is the *conditional entropy* of the output of channel V given an input of distribution P. For such a P and V we denote by PV the distribution on $\mathcal{Y}$, obtained by

$$PV(y) \triangleq \sum_{x \in \mathcal{X}} P(x)\, V(y|x) ;$$

this is the output distribution of channel V corresponding to an input of distribution P. The quantity

$$I(P,V) \triangleq H(PV) - H(V|P) = \sum_{x \in \mathcal{X}} \sum_{y \in \mathcal{Y}} P(x)\, V(y|x) \log \frac{V(y|x)}{PV(y)}$$

is called ***mutual information;*** it is a measure of correlation between an input of distribution P and the corresponding output of channel V. Finally, for two distributions, S and T on $\mathcal{Y}$, we denote by $D(S \| T)$ their *informational divergence* defined as

$$D(S\|T) \triangleq \sum_{y \in \mathcal{Y}} S(y) \log \frac{S(y)}{T(y)} ,$$

and by $D(V\|W|P)$ we denote the *average informational divergence* of the rows of the stochastic matrices $V: \mathcal{X} \to \mathcal{Y}$ and $W: \mathcal{X} \to \mathcal{Y}$ corresponding to the averaging distribution P on $\mathcal{X}$, i.e.

$$D(V\|W|P) \triangleq \sum_{x \in \mathcal{X}} P(x)\, D(V(\cdot|x) \,\|\, W(\cdot|x)) =$$

$$= \sum_{x \in \mathcal{X}} P(x) \sum_{y \in \mathcal{Y}} V(y|x) \log \frac{V(y|x)}{W(y|x)} .$$

One easily sees that $D(S\|T)=0$ iff $S(y)=T(y)$ for every $y\in\mathcal{Y}$, i.e. iff the two distributions coincide.

We shall also speak of the information content of RV's. The *conditional entropy* of a RV Y given the RV X is

$$H(Y|X) \triangleq H(P_{Y|X}\,|P_X)$$

where P_X denotes the distribution of X and $P_{Y|X}$ is the stochastic matrix of the conditional distributions of Y given the various values of X. The *entropy* of the RV Y is the conditional entropy of Y given any constant-valued RV;

$$H(Y) \triangleq H(P_Y).$$

The *mutual information* of the RV's X and Y is

$$I(X\wedge Y) \triangleq H(Y|X) - H(Y) = I(P_X, P_{Y|X}).$$

Finally, the *conditional mutual information* of the RV's X and Y given the RV Z is

$$I(X\wedge Y|Z) \triangleq H(Y|Z) - H(Y|ZX).$$

Recall the following notation: $\lfloor t \rfloor$ is the largest integer not exceeding t, and $|t|^+ \triangleq \max(t,0)$. Throughout the paper, exp's and log's are to the base 2.

In our constructions we shall be concerned with sequences of elements of the alphabets $\mathcal{X}$ and $\mathcal{Y}$. The property of a sequence playing a decisive role in the sequel is its type. For an arbitrary sequence $\mathbf{x}\in\mathcal{X}^n$ and element $x\in\mathcal{X}$ we denote by $N(x|\mathbf{x})$ the number of those coordinates i $(1\leq i\leq n)$ for which $\mathbf{x}=x_1x_2\dots x_n$ satisfies $x_i=x$. Clearly, the numbers $N(x|\mathbf{x})$, $x\in\mathcal{X}$ satisfy

$$\sum_{x\in\mathcal{X}} N(x|\mathbf{x}) = n,$$

and hence $\{1/n\, N(x|\mathbf{x})\}_{x\in\mathcal{X}}$ defines a probability distribution on $\mathcal{X}$ called the *type* of $\mathbf{x}$. We denote this distribution by $P_{\mathbf{x}}$. Similarly, the joint distribution on $\mathcal{X}\times\mathcal{Y}$ defined by the sequence $(\mathbf{x},\mathbf{y})$, considered as a sequence of elements of $\mathcal{X}\times\mathcal{Y}$ is called the *joint type* of $\mathbf{x}$ and $\mathbf{y}$, denoted by $P_{\mathbf{x},\mathbf{y}}$.

For every integer n let $\mathcal{P}_n$ denote the family of those distributions on $\mathcal{X}$ which occur as the type of some sequence $\mathbf{x}\in\mathcal{X}^n$. Further, for $P\in\mathcal{P}_n$ denote by $\mathcal{V}_n(P)$ the family of those stochastic matrices $V:\mathcal{X}\to\mathcal{Y}$ which for some $\mathbf{x}\in\mathcal{X}^n$ and $\mathbf{y}\in\mathcal{Y}^n$ yield

$$P_{\mathbf{x},\mathbf{y}}(x,y) = P_{\mathbf{x}}(x)V(y|x) \quad \text{for every} \quad x\in\mathcal{X}, \ y\in\mathcal{Y}. \tag{10}$$

If **x,y** satisfy this relation for a stochastic matrix V we shall say that **y** has *conditional type* V *given* **x**. This conditional type of a sequence is not uniquely determined; namely if $P_{\mathbf{x}}(x) = 0$ for some $x\in\mathcal{X}$ then any distribution $V(\cdot \mid x)$ satisfies (10).

We denote by $\mathcal{T}_P$ the set of those sequences $\mathbf{x}\in\mathcal{X}^n$ which have type P, and by $\mathcal{T}_V(\mathbf{x})$ the set of those sequences $\mathbf{y}\in\mathcal{Y}^n$ which have conditional type $V:\mathcal{X}\to\mathcal{Y}$ given $\mathbf{x}\in\mathcal{X}^n$. The following elementary lemmas feature information quantities as asymptotic exponents of the polynomial coefficients describing the cardinality of the sets $\mathcal{T}_P$ and $\mathcal{T}_V(\mathbf{x})$.

First of all, observe that "there are not many types":

Lemma 1. For every integer n

$$|\mathcal{P}_n| < (n+1)^{|\mathcal{X}|}$$

and

$$|\mathcal{V}_n| < (n+1)^{|\mathcal{X}||\mathcal{Y}|},$$

where in the second inequality $|\mathcal{V}_n|$ is understood in the sense that out of those stochastic matrices which can be conditional types of a fixed sequence **y** given a fixed sequence **x** only one is counted for.

Proof. This lemma is trivial. The second inequality follows from the first one to prove the latter, observe that each element of $\mathcal{P}_n$ is an $|\mathcal{X}|$-dimensional vector of rationals of the form k/n, with $0\leqslant k \leqslant n$. Every coordinate of such a vector can have at most n + 1 different values.

Next observe that for $\mathbf{x}\in\mathcal{T}_P$, $\mathbf{y}\in\mathcal{T}_V(\mathbf{x})$ and an arbitrary stochastic matrix $W:\mathcal{X}\to\mathcal{Y}$, the definitions yield

$$W^n(\mathbf{y}|\mathbf{x}) = \exp\{-n[D(V\|W|P) + H(V|P)]\}.$$

The following lemma exhibits a Stirling-type but non-asymptotic bound on the size of $\mathcal{T}_P$.

Lemma 2. For every $P\in\mathcal{P}_n$

$$(n+1)^{-|\mathcal{X}|}\exp\{nH(P)\} \leqslant |\mathcal{T}_P| \leqslant \exp\{nH(P)\}.$$

Corollary 2. For every $x \in \mathcal{X}$ and $V \in \mathcal{V}_n(P_x)$

$$(n+1)^{-|\mathcal{X}||\mathcal{Y}|} \exp\{nH(V|P_x)\} \leq |\mathcal{T}_V(x)| \leq \exp\{nH(V|P_x)\}.$$

Proof. By (11), for every $x \in \mathcal{T}_P$

$$P^n(x) = \exp\{-nH(P)\},$$

whence

$$1 \geq P^n(\mathcal{T}_P) \doteq |\mathcal{T}_P| \cdot \exp\{-nH(P)\}, \tag{12}$$

yielding the upper bound. To prove the lower bound, we first show that the probability $P_n(\mathcal{T}_Q)$ is maximized for $Q \triangleq P$. In fact,

$$P^n(\mathcal{T}_Q) = n! \prod_{x \in \mathcal{X}} \frac{P(x)^{nQ(x)}}{(nQ(x))!},$$

whence we obtain

$$\frac{P^n(\mathcal{T}_Q)}{P^n(\mathcal{T}_P)} = \prod_{x \in \mathcal{X}} \frac{(nP(x))!}{(nQ(x))!} P(x)^{n[Q(x)-P(x)]}.$$

Applying the obvious inequality

$$\frac{k!}{\ell!} \leq k^{k-\ell}$$

for $k \triangleq nP(x)$, $\ell \triangleq nQ(x)$, the previous relation gives

$$\frac{P^n(\mathcal{T}_Q)}{P^n(\mathcal{T}_P)} \leq \prod_{x \in \mathcal{X}} n^{n[P(x)-Q(x)]} = 1.$$

Combining this with Lemma 1 and (12) proves the missing inequality, since

$$1 = \sum_{Q \in \mathcal{P}_n} P^n(\mathcal{T}_Q) \leq (n+1)^{|\mathcal{X}|} P^n(\mathcal{T}_P) =$$

$$= (n+1)^{|\mathcal{X}|} |\mathcal{T}_P| \cdot \exp\{-nH(P)\}.$$

Now we are well equipped to prove the theorem.

Existence of Good Codes

First we shall derive the promised lower bound on E(R,W). To this end we shall consider block codes of a special type. With every subset $\mathcal{C}$ of $\mathcal{X}^n$ we shall associate a code of block length n as follows: Let the message set $\mathcal{M}$ be identical to $\mathcal{C}$ and let the encoder f be the identity mapping on $\mathcal{C}$.

In order to define the decoder we need a new notation. For a pair of sequences $\mathbf{x} \in \mathcal{X}^n$, $\mathbf{y} \in \mathcal{Y}^n$, let us denote by $I(\mathbf{x} \wedge \mathbf{y})$ the mutual information $I(P_\mathbf{x}, V)$ where V is the conditional type of **y** given **x**. Although the conditional type of **y** given **x** is not uniquely determined, nonetheless all the involved stochastic matrices V yield the same mutual information $I(P_\mathbf{x},V)$ and this $I(\mathbf{x} \wedge \mathbf{y})$ is well-defined. Now a function $\varphi: \mathcal{Y}^n \to \mathcal{C}$ will be called a *maximum mutual information* (MMI) *decoder* associated with $\mathcal{C}$ if it has the property that for every $\mathbf{y} \in \mathcal{Y}^n$ the sequence $\varphi(\mathbf{y}) \in \mathcal{C}$ maximizes $I(\mathbf{x} \wedge \mathbf{y})$ as **x** ranges over the elements of $\mathcal{C}$, i.e.

$$I(\varphi(\mathbf{y}) \wedge \mathbf{y}) = \max_{\mathbf{x} \in \mathcal{X}} I(\mathbf{x} \wedge \mathbf{y}) . \tag{13}$$

Any MMI-decoder associated with $\mathcal{C}$ will equally suit our purposes. (We should mention that these decoders have been introduced by Goppa [14]. For the justification of this apparently very peculiar choice of the decoder cf. Csiszár-Körner [9].)

Anything else being settled in advance, we shall complete our choice of block codes of length n by specifying the sets $\mathcal{C}$. This means that once n and the rate R have been fixed, a universal set $\mathcal{C} \subset \mathcal{X}^n$ is chosen to define a block code of prescribed block length and rate in the above manner. We shall return to the significance of this circumstance later.

The construction of $\mathcal{C}$ is based on a special case of an elementary combinatorial lemma from Csiszár-Körner [9] which in turn is a continuous version of a graph decomposition result of Lovász [15].

Lemma 3. Let $\mathcal{A}$ be a finite set and let ν be a non-negative function on $\mathcal{A} \times \mathcal{A}$ with the properties that

(i) $\nu(a,b) = \nu(b,a)$ for every $a \in \mathcal{A}$.

(ii) $\nu(a,a) = 0$ for every $a \in \mathcal{A}$.

If d is any number satisfying

$$\sum_{b \in \mathcal{A}} \nu(a,b) < d \quad \text{for every} \quad a \in \mathcal{A} , \tag{14}$$

then for every positive integer s and real number t with

(15) $$st \geq d$$

the set $\mathcal{A}$ has a partition into disjoint subsets $\mathcal{A}_1, \ldots, \mathcal{A}_s$ so that

$$\sum_{b \in \mathcal{A}_i} \nu(a,b) < t \quad \text{for every} \quad 1 \leq i \leq s \quad \text{and} \quad a \in \mathcal{A}_i .$$

Corollary 3. Let $\mathcal{A}$ be a finite set and let ν be a function as in the lemma. If d satisfies (14) and s, t are arbitrary real numbers satisfying (15), then there is a set $\mathcal{A}_0 \subset \mathcal{A}$ such that

(16) $$|\mathcal{A}_0| \geq \frac{|\mathcal{A}|}{s}$$

and

$$\sum_{b \in \mathcal{A}_0} \nu(a,b) < t \quad \text{for every} \quad a \in \mathcal{A}_0 .$$

Proof. Fixing some s and t satisfying (15), let $\mathcal{A}_1, \ldots, \mathcal{A}_s$ be the classes of any partition of $\mathcal{A}$ such that the sum of the "inside values of ν" i.e.

(17) $$\sum_{i=1} \sum_{a \in \mathcal{A}_i} \sum_{b \in \mathcal{A}_i} \nu(a,b)$$

is minimum. We claim that this partition has the needed properties. In fact, the minimizing property means that for every i and $a \in \mathcal{A}_i$, moving a into any $\mathcal{A}_j$, the sum (17) can only increase, i.e.

$$\sum_{b \in \mathcal{A}_j} \nu(a,b) - \sum_{b \in \mathcal{A}_i} \nu(a,b) \geq 0 .$$

Summing these inequalities for $j = 1,2, \ldots, s$ we get

$$\sum_{b \in \mathcal{A}} \nu(a,b) = \sum_{j=1}^{s} \sum_{b \in \mathcal{A}_j} \nu(a,b) \geq \sum_{b \in \mathcal{A}_i} \nu(a,b) .$$

Hence

$$\sum_{b \in \mathcal{A}_i} \nu(a,b) \leq \frac{1}{s} \sum_{b \in \mathcal{A}} \nu(a,b) < \frac{d}{s} ,$$

whereby the lemma follows from (15).

The corollary is established by letting $\mathcal{A}_0$ be any $\mathcal{A}_i$ of maximum size.

A straightforward application of Corollary 3 will provide us with the set $\mathcal{C}$. This is the content of the next lemma of [9] which was originally proved in a more complicated manner in [12].

Lemma 4. For any finite set $\mathcal{X}$, distribution $P \in \mathcal{P}_n$ and positive number

$$r < |\mathcal{T}_P|$$

there exists a set $\mathcal{C} \subset \mathcal{T}_P$ of size

$$|\mathcal{C}| \geqslant r$$

such that for every $\mathbf{x} \in \mathcal{C}$ and every stochastic matrix $\bar{V}: \mathcal{X} \to \mathcal{X}$ different from the identity one has

$$|\mathcal{T}_{\bar{V}}(\mathbf{x}) \cap \mathcal{C}| \leqslant r|\mathcal{T}_{\bar{V}}(\mathbf{x})| \exp\{-n(H(P) - \delta_n)\},$$

where the sequence $\delta_n \to 0$ is given by

$$\delta_n \triangleq (|\mathcal{X}|^2 + |\mathcal{X}|) \frac{\log(n+1)}{n} + \frac{1}{n}.$$

Proof. Choose $\mathcal{A} \triangleq \mathcal{T}_P$ and apply Corollary 3 to this set and a function ν defined as

$$\nu(\mathbf{x},\mathbf{x}) \triangleq \begin{cases} \dfrac{1}{|\mathcal{T}_{\bar{V}}(\mathbf{x})|} & \text{if } \tilde{\mathbf{x}} \neq \mathbf{x}, \quad \tilde{\mathbf{x}} \in \mathcal{T}_{\bar{V}}(\mathbf{x}) \\ 0 & \text{if } \tilde{\mathbf{x}} = \mathbf{x}. \end{cases}$$

Check first that ν has the required properties. To this end notice that although the conditional type $\bar{V}$ of $\tilde{\mathbf{x}}$ given $\mathbf{x}$ is not uniquely determined, $|\mathcal{T}_{\bar{V}}(\mathbf{x})|$ still is. The symmetry of ν follows by observing that if $\bar{\bar{V}}$ is the conditional type of $\mathbf{x}$ given $\tilde{\mathbf{x}}$, then

$$|\mathcal{T}_P| \cdot |\mathcal{T}_{\bar{V}}(\mathbf{x})| = |\mathcal{T}_P| \cdot |\mathcal{T}_{\bar{\bar{V}}}(\tilde{\mathbf{x}})|,$$

as both sides equal the number of those ordered pairs of sequences from $\mathcal{X}^n$ which have the same joint type as $(\mathbf{x},\tilde{\mathbf{x}})$.

It follows from Lemma 1 that for every $x \in \mathcal{T}_P$

$$\sum_{\tilde{x} \in \mathcal{T}_P} \nu(x, \tilde{x}) = \sum_{\bar{V} \in \mathcal{V}_n(P)} 1 < (n+1)^{|\mathcal{X}|^2} .$$

Choosing $d \triangleq (n+1)^{|\mathcal{X}|^2}$ and $s \triangleq \lfloor 1/r |\mathcal{T}_P| \rfloor$ the lemma follows from Corollary 3.

Now, the direct part of the Noisy Channel Coding Theorem follows. More importantly, we shall establish the random coding bound of Fano-Gallager, ([10], [11]).

Theorem 1. For every type P of sequences in $\mathcal{X}^n$ and $R > 0$ there exists a set $\mathcal{C} \subset \mathcal{T}_P$ with

$$\frac{1}{n} \log |\mathcal{C}| \geqslant R - \delta_n$$

such that for a DMC given by an arbitrary stochastic matrix $W: \mathcal{X} \to \mathcal{Y}$ the code (f, φ) associated with the codeword set $\mathcal{C}$ yields average probability of error

$$\bar{e}(W^n, f, \varphi) \leqslant \exp\{-n(E_r(R, P, W) - \delta_n')\} ,$$

where

$$E_r(R, P, W) \triangleq \min_V D(V||W|P) + |I(P, V) - R|^+ ,$$

$$\delta_n \triangleq (|\mathcal{X}|^2 + |\mathcal{X}|) \frac{\log(n+1)}{n} + \frac{1}{n} ,$$

$$\delta_n' \triangleq |\mathcal{X}|^2 |\mathcal{Y}| \frac{\log(n+1)}{n} .$$

Proof: Let $\mathcal{C}$ be the set of Lemma 4 corresponding to $r \triangleq \exp\{n(R - \delta_n)\}$. By Lemma 4 and Corollary 2, for every $x \in \mathcal{C}$ and every $V: \mathcal{X} \to \mathcal{X}$ different from the identity matrix we have

$$|\mathcal{T}_{\bar{V}}(x) \cap \mathcal{C}| \leqslant \exp\{n(R - I(P, \bar{V})\} . \tag{18}$$

Let us now consider those sequences y for which the MMI-decoder associated with $\mathcal{C}$ commits an error. By (13), every such y is contained in the set $\mathcal{E}(x)$ of those sequences y for which there is another $\hat{x} \in \mathcal{C}$ different from x with the property

$$I(\hat{x} \wedge y) \geqslant I(x \wedge y) . \tag{19}$$

Clearly,

$$e(W^n, f, \varphi) \leq \max_{x \in \mathcal{C}} W^n(\mathcal{E}(x)|x) . \tag{20}$$

We shall distinguish between the elements of $\mathcal{E}(x)$ according to the joint type $P_{x\hat{x}y}$ of the triple $(x,\hat{x},y)$ involved in (19). Let us denote by $\mathcal{E}(x,Q)$ the set of those sequences $y \in \mathcal{E}(x)$ for which (19) holds with some $\hat{x} \in \mathcal{C}$ such that

$$P_{x\hat{x}y} = Q . \tag{21}$$

By Lemma 1, there are at most $(n+1)^{|\mathcal{X}|^2|\mathcal{Y}|}$ different distributions on $|\mathcal{X}|^2|\mathcal{Y}|$ which occur as such joint types. Hence

$$W^n(\mathcal{E}(x)|x) \leq (n+1)^{|\mathcal{X}|^2|\mathcal{Y}|} \max_Q W^n(\mathcal{E}(x, Q)|x) , \tag{22}$$

where Q ranges over distributions satisfying (21) and (19) for some $\hat{x} \in \mathcal{C}$ and $y \in \mathcal{Y}^n$.

For convenience, let $X, \hat{X}$ and Y denote an arbitrary triple of RV's with a joint distribution Q as above. Then the conditions (19), (21) and $x \in \mathcal{C}$, $\hat{x} \in \mathcal{C}$ imply

$$P_X = P_{\hat{X}} = P , \qquad I(\hat{X} \wedge Y) \geq I(X \wedge Y) . \tag{23}$$

Thus (22) can be further bounded as

$$W^n(\mathcal{E}(x)|x) \leq (n+1)^{|\mathcal{X}|^2|\mathcal{Y}|} \max_{P_{X\hat{X}Y}} W^n(\mathcal{E}(x, P_{X\hat{X}Y})|x) , \tag{24}$$

with the maximization being extended to arbitrary triples of RV's satisfying (23).

As all the elements of $\mathcal{E}(x, P_{X\hat{X}Y})$ have the same conditional type given x, we can compute the right-hand side of (24) by bounding the cardinality of $\mathcal{E}(x, P_{X\hat{X}Y})$. We shall bound the number of those sequences **y** which satisfy (19) for the fixed $x \in \mathcal{C}$ and some $\hat{x} \in \mathcal{C}$ such that $P_{x\hat{x}y} = P_{x\hat{x}y}$ by the number of the pairs $(y\hat{x})$ with the same properties. This yields, using Corollary 2 and denoting $V \triangleq P_{\hat{X}|X}$,

$$|\mathcal{E}(x, P_{X\hat{X}Y})| \leq \exp\{n H(Y|X\hat{X})\} \cdot |\mathcal{T}_V(x) \cap \mathcal{C}| ,$$

whence by property (18) of the set $\mathcal{C}$ we obtain

$$|\mathcal{E}(x, P_{X\hat{X}Y})| \leq \exp\{ n[H(Y|X\hat{X}) + R - I(X \wedge \hat{X})]\} . \tag{25}$$

Observe now that by Corollary 2 the obvious bound

$$(26) \qquad |\mathcal{E}(\mathbf{x}, P_{X\hat{X}Y})| \leq \exp\{n H(Y|X)\}$$

necessarily holds true. In order to combine the last two inequalities, recall that by definition,

$$H(Y|X\hat{X}) = H(Y|X) - I(Y \wedge \hat{X}|X) .$$

Substituting this into (25) and taking into account (26) we get

$$|\mathcal{E}(\mathbf{x}, P_{X\hat{X}Y})| \leq \exp\{n[H|(Y|X) - |I(YX \wedge \hat{X}) - R|^{+}]\} .$$

Here, using (23), we see that

$$I(YX \wedge \hat{X}) \geq I(Y \wedge \hat{X}) \geq I(Y \wedge X) ,$$

yielding

$$|\mathcal{E}(\mathbf{x}, P_{X\hat{X}Y})| \leq \exp\{n[H(Y|X) - |I(Y \wedge X) - R|^{+}]\} .$$

Denoting $P_{Y|X}$ by $V: \mathcal{X} \to \mathcal{Y}$ and using (11) the last relation becomes

$$|\mathcal{E}(\mathbf{x}, P_{X\hat{X}Y})| \leq \exp\{-n[D(V||W|P) + I|(P, V) - R|^{+}]\} .$$

Comparing this with (20) and (24) we can complete the proof.

The last theorem immediately gives

Theorem 2. For every stochastic matrix $W: \mathcal{X} \to \mathcal{Y}$ and $R \geq 0$

$$E(R, W) \geq \max_{P} E_r(R, P, W) =$$

$$= \max_{P} \min_{V} D(V||W|P) + |I(P, V) - R|^{+} .$$

Proof. Every distribution P on $\mathcal{X}$ can be approximated by elements of $\mathcal{P}_n$ as n tends to infinity. Apply Theorem 1 to a sequence of types approximating the distribution P achieving the maximum of E_r (R,P,W).

We shall discuss our existence results later. Before doing so, let us turn to

Converse Results

These results will be far more easy to obtain. The essence of the proof of Dueck-Körner [8] is the following simple observation:

Lemma 5. For any $R \geq 0$, $P \in \mathcal{P}_n$, stochastic matrix $V: \mathcal{X} \to \mathcal{Y}$, every collection of not necessarily distinct sequences $\mathbf{x}_i$, $1 \leq i \leq M$ such that

$$\mathbf{x}_i \in \mathcal{T}_P, \qquad M \geq (n+1)^{|\mathcal{X}||\mathcal{Y}|} \exp\{nR\} \tag{27}$$

and every mapping $\varphi: \mathcal{Y}^n \to \{\mathbf{x}_1, \ldots, \mathbf{x}_M\}$ we have

$$\frac{1}{M} \sum_{1 \leq i \leq M} \frac{|\mathcal{T}_V(\mathbf{x}_i) \cap \varphi^{-1}(i)|}{|\mathcal{T}_V(\mathbf{x}_i)|} \leq \exp\{-n|R - I(P,V)|^+\}, \tag{28}$$

provided that $V \in \mathcal{V}_n(P)$.

Proof. It is enough to show that the left-hand side of (28) does not exceed $\exp\{-n[R-I(P,V)]\}$, as it can never exceed 1. By Corollary 2, the left-hand side of (28) is at most

$$\frac{1}{M}(n+1)^{|\mathcal{X}||\mathcal{Y}|} \exp\{-n H(V|P)\} \sum_{1 \leq i \leq M} |\mathcal{T}_V(\mathbf{x}_i) \cap \varphi^{-1}(i)| \leq$$

$$\leq \exp\{-n[R + H(V|P)]\} \sum_{1 \leq i \leq M} |\mathcal{T}_V(\mathbf{x}_i) \cap \varphi^{-1}(i)|.$$

Noticing that the sets $\mathcal{T}_V(\mathbf{x}_i) \cap \varphi^{-1}(i)$ are disjoint subsets of $\mathcal{T}_{PV}$ for different values of i and using the symmetry of mutual information, the last expression can be upper bounded according to Lemma 2 by $\exp\{-n[R - I(P,V)]\}$ which is the needed bound.

The final converse result is obtained from this lemma in a routine manner.

Theorem 3. For every $R \geq 0$, every block code (f, φ) of block length n and rate

$$R_f \geq R + |\mathcal{X}||\mathcal{Y}| \frac{\log(n+1)}{n}$$

and every stochastic matrix $W: \mathcal{X} \to \mathcal{Y}$ the average probability of error of (f, φ), when applied to the DMC given by W, satisfies

$$1 - \bar{e}(W^n, f, \varphi) \leq \exp\{-n[\hat{G}(R,W) - \delta''_n]\} \tag{29}$$

where

$$\hat{G}(R,W) = \min_P \min_V [D(V\|W|P) + |R - I(P,V)|^+]$$

and

$$\delta''_n \triangleq (|\mathcal{X}| + |\mathcal{X}||\mathcal{Y}|) \frac{\log(n+1)}{n} .$$

Proof. We claim that for some type $P \in \mathcal{P}_n$ of sequences there is a new code $(\hat{f}, \hat{\varphi})$ with codewords of type P such that denoting the average probability of correct decoding by

$$\bar{c}(W^n, f, \varphi) \triangleq 1 - \bar{e}(W^n, f, \varphi) ,$$

we have

$$\bar{c}(W^n, f, \varphi) \leq (n+1)^{|\mathcal{X}|} \bar{c}(W^n, \hat{f}, \hat{\varphi}) \tag{30}$$

while the message set $\mathcal{M}$ remains the same. In fact, let P be any element of $\mathcal{P}_n$ which maximizes

$$\frac{1}{|\mathcal{M}|} \sum_{m : f(m) \in \mathcal{T}_P} W^n(\varphi^{-1}(m)|f(m)) .$$

Define

$$\hat{f}(m) \triangleq \begin{cases} f(m) & \text{if} \quad f(m) \in \mathcal{T}_P \\ \text{arbitrary element of } \mathcal{T}_P & \text{else} , \end{cases}$$

$$\hat{\varphi}(m) \triangleq \varphi(m) \qquad m \in \mathcal{M}$$

Obviously, this code $(\hat{f}, \hat{\varphi})$ satisfies our claim. Furthermore, we have

$$\bar{c}(W^n, \hat{f}, \hat{\varphi}) \triangleq \frac{1}{|\mathcal{M}|} \sum_{m \in \mathcal{M}} W^n(\hat{\varphi}^{-1}(m)|f(m)) \leq$$

$$\leq \sum_{V \in \mathcal{V}_n(P)} \frac{1}{|\mathcal{M}|} \sum_{m \in \mathcal{M}} W^n(\hat{\varphi}^{-1}(m) \cap \mathcal{T}_V(f(m))|f(m)) =$$

$$= \sum_{V \in \mathcal{V}_n(P)} \frac{1}{|\mathcal{M}|} \sum_{m \in \mathcal{M}} W^n(\mathcal{T}_V(f(m))|f(m)) \cdot \frac{|\mathcal{T}_V(f(m)) \cap \varphi^{-1}(m)|}{|\mathcal{T}_V(f(m))}$$

$$= \sum_{V \in \mathcal{V}_n(P)} \exp\{-nD(V||W|P)\} \cdot \frac{1}{|\mathcal{M}|} \sum_{m \in \mathcal{M}} \frac{|\mathcal{T}_V(f(m)) \cap \varphi^{-1}(m)|}{\mathcal{T}_V(f(m))} ,$$

with the last inequality following by Corollary 2 and (11). Thus, using the last lemma we conclude that

$$\tau(W^n, \hat{f}, \hat{\varphi}) \leqslant \sum_{V \in \mathcal{V}_n(P)} \exp\{-n[D(V\|W|P) + |R - I(P,V)|^+]\} .$$

By Lemma 1 this yields

$$\bar{\tau}(W^n, \hat{f}, \hat{\varphi}) \leqslant (n+1)^{|\mathscr{X}||\mathscr{Y}|} \exp\{-n \min_V D(V\|W|P) + |R - I(P,V)|^+\} ,$$

and therefore, a fortiori,

$$c(W^n, \hat{f}, \hat{\varphi}) \leqslant (n+1)^{|\mathscr{X}||\mathscr{Y}|} \exp\{-n\, \hat{G}(R, W)\} .$$

Comparing this inequality with (30) we can prove the theorem.

Hence we immediately obtain the counterpart of Theorem 2:

Theorem 4. For every stochastic matrix $W : \mathscr{X} \to \mathscr{Y}$ and $R \geqslant 0$

$$G(R, W) \geqslant \hat{G}(R, W) = \min_P \min_V D(V\|W|P) + |R - I(P,V)|^+ .$$

Proof. Obvious.

It is noticeable how similar the two bounds are.

At this point time has come to analyse the results and derive the Noisy Channel Coding Theorem.

Conclusions

The results of Theorems 1 and 3 are non-trivial only when $E_r(R,P,W)$ and $\hat{G}(R,W)$ are positive. This motivates our interest in

Lemma 6.

$$E_r(R, P, W) > 0 \quad \text{iff} \quad I(P, W) > R .$$

$$\hat{G}(R, W) > 0 \quad \text{iff} \quad \max_P I(P, W) < R .$$

Proof. It is clear that $R \geqslant I(P,W)$ implies

$$E_r(R, P, W) \leqslant D(W\|W|P) + |I(P,W) - R|^+ = 0 .$$

On the other hand, $E_r(R,P,W) = 0$ implies that the minimizing stochastic matrix

$V: \mathcal{X} \to \mathcal{Y}$ satisfies

$$D(V\|W|P) = 0 \tag{31}$$

and

$$|I(P, V) - R|^+ = 0 .$$

Here (31) means that for every $x \in \mathcal{X}$ such that $P(x) > 0$ we have $V(\cdot|x) = W(\cdot|x)$. Thus $I(P,V) = I(P,W)$ and therefore

$$|I(P, W) - R|^+ = |I(P, V) - R|^+ = 0 ,$$

completing the proof of the first statement of the lemma. To prove the second statement, observe that

$$R \leqslant \max_P I(P, W)$$

implies for every distribution P maximizing I(P,W) that

$$\hat{G}(R, W) \leqslant D(W\|W|P) + |R - I(P, W)|^+ = 0 .$$

Finally, $\hat{G}(R,W) = 0$ implies that for every minimizing pair P and V relation (31) holds, whereby $I(P,V) = I(P,W)$, as noticed earlier. As for the minimizing pair we also have $|R-I(P,V)|^+ = 0$, the proof is complete.

The above lemma leads us to the

Noisy Channel Coding Theorem. For every $\varepsilon \in (0,1)$, the ε-capacity C_ε of the DMC given by $W: \mathcal{X} \to \mathcal{Y}$ is

$$C_\varepsilon = \max_P I(P, W) .$$

Proof. Lemma 6 shows that $C_\varepsilon \geqslant \max_P I(P, W)$, since for $R < \max_P I(P, W)$ we have $\max_P E_r(R,P,W) > 0$, and by Theorem 2 the latter implies $E(R,W) > 0$.

On the other hand, the previous lemma also shows that for $R > \max_P I(P,W)$ we have $\hat{G}(R,W) > 0$. By Theorem 4 this means that $G(R,W) > 0$, and thus

$$C_\varepsilon \leqslant \max_P I(P, W) .$$

We have established the promised lower bounds on E(R,W) and G(R,W), thus proving the coding theorem with strong converse for DMC's with finite alphabets. Let us dwell for a moment on the problem of error exponents.

While, as shown by Duek and Körner [8], G(R,W) equals $\hat{G}(R,W)$, the determination of E(R,W) is a formidably difficult open problem. Shannon, Gallager and Berlekamp [16] showed that

$$E(R, W) \neq \max_P E_r(R, P, W)$$

although the two sides are actually equal for every W in a neighbourhood of the capacity of the DMC given by W. The left endpoint of this neighbourhood can be easily computed; it is different from capacity except for trivial cases, and has been called the *critical rate* of the DMC.

Gallager [11] has improved on the lower bound

$$E(R, W) \geqslant \max_P E_r(R, P, W)$$

by introducing his expurgated bound. The interested reader is advised to consult [9] where this bound is derived to the analogy of Theorem 1, using the combinatorial approach and the very same Lemma 4. Actually, one establishes the expurgated bound by constructing a code with the same codeword set $\mathcal{C}$ as here, but using a maximum likelihood decoder instead of MMI-decoders.

The main concern of the celebrated paper of Shannon-Gallager-Berlekamp [16] was to establish upper bounds on E(R,W). They did this in a rather complicated manner. Their bound, called the *sphere-packing bound,* was rederived later by a simple and elegant proof, independently by Haroutunian [17] and Blahut [18]. Haroutunian and Blahut showed that

$$E(R, W) \geqslant E_{sp}(R, W) \triangleq \max_P \min_{V:I(P,V) \leqslant R} D(V\|W|P) ;$$

also this form of the bound is new.

Let us add that E(R,W) has been determined for $\overline{R} = 0$ in Shannon-Gallager-Berlekamp [16] where several other interesting results can be found.

We have mentioned that maximum-likelihood decoding is superior to MMI-decoding. In fact, for a given encoder maximum likelihood decoding rules guarantee the smallest average probability of error. Yet, they have a serious practical disadvantage.

Universal Coding

Maximum likelihood decoders can be applied only if the statistics of the channel is precisely known. This, however is often impossible in practice. It is natural to ask therefore what can be done without involving the channel statistics. There are several mathematical formulations of this problem. Without explaining their meaning, we should mention compound channels, stochastically varying channels or arbitrarily varying channels. (The interested reader can consult [7] or the recent book [19].) In all the mentioned models one supposes that the transition probabilities of the channel form a stochastic matrix which belongs to a known set of matrices.

A more ambitious way to approach this practical problem is offered by universal coding. This concept was invented in a source coding context by Fitingof [20], Lynch [21] and Davisson [22]. For a large class of sources, involving all the discrete memoryless sources with the same alphabet, universally optimal codes exist in some asymptotic sense. This means that for every block length, a unique code has a performance which is uniformly close to the optimum for each source within the class. For more on this, cf. e.g. Davisson [23]. Universal optimality in this sense is no more achievable even for the class of all DMC's with the same input and output alphabets.

In a broader sense, universal coding means to measure the goodness of codes by using performance criteria relative to a whole class of systems (sources or channels). In our case it will mean to measure the goodness of block codes for the class of all DMC's with given alphabets by evaluating the whole spectrum of their error probabilities for each DMC in the class.

Let us denote by $\mathcal{W}(\mathcal{X}\to\mathcal{Y})$ the family of all stochastic matrices of which the $|\mathcal{X}|$ rows are distributions of $\mathcal{Y}$, each row being indexed by a different element of $\mathcal{X}$. We shall say that a non-negative function E with domain $\mathcal{W}(\mathcal{X}\to\mathcal{Y})$ is a *universally attainable error exponent* UEX) *at rate* $R\geqslant 0$ for the class of DMC's with the given alphabets, if for every $\delta>0$ and sufficiently large n there exists a single block code (f,φ) of block length n and rate

$$R_f \geqslant R-\delta$$

such that

$$e(W^n, f, \varphi) \leqslant \exp\{-n[E(W)-\delta]\} \quad \text{for every } W\in\mathcal{W}(\mathcal{X}\to\mathcal{Y}).$$

Of course, there are trivial examples of such universally attainable error exponents. One can say, e.g. that fixing some DMC $W_0\in\mathcal{W}(\mathcal{X}\to\mathcal{Y})$, the function

$$\hat{E}(W) \triangleq \begin{cases} E(R, W_0) & \text{if} \quad W = W_0 \\ 0 & \text{else} \end{cases}$$

is a universally attainable error exponent at rate R. Clearly, the relevant question to be asked is to determine the maximal UEX's. A function t_0 on $\mathcal{W}(\mathcal{X} \to \mathcal{Y})$ is a *maximal UEX at rate* $R \geq 0$, if every other UEX $\tilde{E}$ at rate R satisfies

$$\tilde{E}(W) < E_0(W)$$

at least for one $W \in \mathcal{W}(\mathcal{X} \to \mathcal{Y})$. The determination of all these maximal UEX's is even harder than that of reliability functions, the latter being involved in the former. In Csiszár-Körner-Marton [12] the following non-trivial example was given:

Theorem 5. For every distribution P on and $R \geq 0$ the function

$$E(W) \triangleq E_r(R, P, W) , \qquad W \in \mathcal{W}(\mathcal{X} \to \mathcal{Y})$$

is a UEX at rate R for the class of DMC's with the given alphabets.

Proof. Observe that in Theorem 1 neither the encoder nor the decoder were chosen depending on W.

The last theorem yields a different UEX at rate R for every distribution P on $\mathcal{X}$. These functions do not compare. Csiszár and Körner [9] have considerably improved on Theorem 5, thus showing that the UEX's of this thoerem are not maximal. Their improved bound was obtained by the method presented here, and all what was needed was more careful computation.

We do not want to enter these technical details. Let us just mention one of our hitherto neglections, as the underlying observation merits independent attention. We state an improved version of Lemma 4.

Lemma 7 . For every type P of sequences in $\mathcal{X}^n$ and

$$0 < R < H(P)$$

there exists a set $\mathcal{C} \subset \mathcal{T}_P$ with

$$\frac{1}{n} \log |\mathcal{C}| > R - \delta_n \tag{32}$$

such that for every $x \in \mathcal{C}$, $\hat{x} \in \mathcal{C}$ we have

$$I(x \wedge \hat{x}) \leq R , \tag{33}$$

and also, (18) holds. Here

$$\delta_n \triangleq (|\mathcal{X}|^2 + |\mathcal{X}|) \frac{\log(n+1)}{n} + \frac{1}{n}.$$

Proof. Apply Lemma 1 as in Theorem 1. Then you can construct a set $\mathcal{C}$ satisfying (32) and (18). The additional property (23) is immediate upon observing that for matrices $V: \mathcal{X} \to \mathcal{X}$ with $I(P,V) > R$ (18) implies

$$|\mathcal{T}_V(x) \cap \mathcal{C}| < 1 \qquad \text{for every} \quad x \in \mathcal{C}.$$

This lemma generalizes Gilbert's bound [24] for the minimum Hamming distance in a codeword set. Sets with property (33) were first constructed by Blahut [25]. It is still unknown, whether Gilbert's bound is optimal, cf. McElice-Rodemich-Rumsey-Welch [26].

Outlook

In the field of two terminal communication almost all the information theoretic problems are either solved, or tremendously difficult, at least in the discrete case. In recent years, information theory research was concentrated elsewhere. Much attention was devoted to communication between several terminals. A comprehensive survey on the channel coding problems of this kind is van der Meulen's paper [27]. For the source coding aspects, cf. Csiszár-Körner [28]. Let us briefly mention two applications of the combinatorial approach to channel coding problems of this kind. In [29], Körner and Sgarro obtained universally attainable error exponents for the broadcast channel with degraded message sets. Applying the method, such results can be obtained for several channel models, e.g. the multi-access channel with average probability of error, the arbitrarily varying channel with average probability of error, etc. These results concern error exponents for channels of which the capacity region has been obtained previously. Recently, Csiszár and Körner [30] were able to apply the method so as to improve on a result of Ahlswede [31] about the maximum probability of error capacity of the arbitrarily varying channel. (The determination of the capacity of arbitrarily varying channels is a zero-error type problem for the multi-access channel.)

It was not our aim to enter the various problems of multi-terminal communication. Rather, we wanted to call the reader's attention to a method that has already been successfully applied in this more recent field of Shannon's information theory.

REFERENCES

[1] Shannon, C.E.: The zero-error capacity of a noisy channel IRE-IT, 2, pp. 8-19, 1956.

[2] Lovász, L.: On the Shannon capacity of a graph. IEEE Trans. Inform. Theory, 25, pp. 1-7, 1979.

[3] Haemers, W.: On some problems of Lovász concerning the Shannon capacity of a graph. IEEE Trans. Inform. Theory, 25, pp. 231-232, 1979.

[4] Shannon, C.E.: A mathematical theory of communication, Bell System Techn. Journal, 27, pp. 379-423, 623-656, 1948.

[5] Shannon, C.E.: Certain results in coding theory for noisy channels, Information and Control 1, pp. 6-25, 1957.

[6] Wolfowitz, J.: The coding of messages subject to chance errors, Illinois J. Math. 1, pp. 591-606, 1957.

[7] Wolfowitz, J.: Coding Theorems of Information Theory, Springer Berlin-Heidelberg, 1961, 3rd edition, 1978.

[8] Dueck, G., Körner, J.: Reliability function of a discrete memoryless channel at rates above capacity, IEEE Trans. Inform. Theory, 25, pp. 82-85, 1979.

[9] Csiszár, I., Körner, J.: Graph decomposition: a new key to coding theorems, IEEE Trans. Inform. Theory, to appear.

[10] Fano, R.M.: Transmission of information, A Statistical Theory of Communications, Wiley, New York-London, 1961.

[11] Gallager, R.G.: A simple derivation of the coding theorem and some applications, IEEE Trans. Inform. Theory 11, pp. 3-18, 1965.

[12] Csiszár, I., Körner, J., Marton, K.: A new look at the error exponent of a discrete memoryless channel, Preprint, Presented at the IEEE Int'l. Symposium on Information Theory, 1977, Cornell Univ., Ithaca, N.Y.

[13] Arimoto, S.: On the converse to the coding theorem for discrete memoryless channels, IEEE Trans. Inform. Theory 19, pp. 357-359, 1973.

[14] Goppa, V.D.: Nonprobabilistic mutual information without memory, Problems of Control and Information Theory (in Russian), 4, pp. 97-102, 1975.

[15] Lovász, L.: On decomposition of graphs, Studia Sci. Math. Hung. I, pp. 237-238, 1966.

[16] Shannon, C.E., Gallager, R.G., Berlekamp, E.R.: Lower bounds to error probability for coding in discrete memoryless channels, I-II, Information and Control, 10, pp. 65-103, 522-552.

[17] Haroutunian, E.A.: Estimates of the error exponent for the semicontinuous memoryless channel (in Russian), Problemi Peredaci Informacii. 4, no. 4, pp. 37-48, 1968.

[18] Blahut, R.e.: Hypothesis testing and information theory, IEEE Trans. Inform. Theory, 20, pp. 405-417, 1974.

[19] Csiszár, I., Körner, J.: Information Theory, Coding Theorems for Discrete Memoryless Systems, Academic Press, to appear.

[20] Fitingof, B.M.: Coding in the case of unknown and changing message statistics (in Russian), Problemi Peredaci Informacii, 2, no. 2, pp. 3-11, 1966.

[21] Lynch, T.J.: Sequence time coding for data compression, Proc. IEEE, 54, pp. 1490-1491, 1966.

[22] Davisson, L.D.: Comments on "sequence time coding for data compression", Proc. IEEE 54, p. 2010, 1966.

[23] Davisson, L.D.: Universal noiseless coding, IEEE Trans. Inform. Theory, 19, pp. 783-796, 1973.

[24] Gilbert, E.N.: A comparison of signalling alphabets, Bell System Techn. J. 31, pp. 504-522.

[25] Blahut, R.E.: Composition bounds for channel block codes, IEEE Trans. Inform. Theory, 23, pp. 656-674.

[26] McEliece, R.J. Rodemich, E.R. Rumsey, H. Jr. Welch, L.R.: New upper bounds on the rate of a code via the Delsarte-McWilliams identities, IEEE Trans. Inform. Theory, 23, pp. 157-166, 1977.

[27] van der Meulen, E.C.: A survey of multi-way channels in information theory: 1961-1976, IEEE Trans. Inform. Theory, 23, pp. 1-37, 1977.

[28] Csiszár, I. Körner, J.: Towards a general theory of source networks, IEEE Trans. Inform. Theory, 26.

[29] Körner, J., Sgarro, A.: Universally attainable error exponents for broadcast channels with degraded message sets, IEEE Trans. Inform. Theory, to appear.

[30] Csiszár, I., Körner, J.: On the capacity of the arbitrarily varying channel for maximum probability of error, submitted to Z.f. Wahrscheinlichkeitsthesrie verw. Geb. 1979.

[31] Ahlswede, R.: A method of coding and an application to arbitrarily varying channels, Preprint 1979.

COLLISION-RESOLUTION ALGORITHMS AND RANDOM-ACCESS COMMUNICATIONS

James L. Massey
Professor of System Science
University of California, Los Angeles
Los Angeles CA, 90024

1. Introduction

Communications engineers have a long acquaintance with the "multiple-access" problem, i.e., the problem of providing the means whereby many senders of information can share a common communications resource. The "classical" solution has been to do some form of *multiplexing* (e.g., time-division multiplexing (TDM) or frequency-division multiplexing (FDM) in order to parcel out the resource equitably among the senders.) A fixed division of the resources, however, becomes inefficient when the requirements of the users vary with time. The classical "fix" for the multiplexing solution is to add some form of *demand-assignment* so that the particular division of resources can be adapted to meet the changing requirements. Such demand-assigned multiplexing techniques have proved their worth in a myriad of multiple-access applications.

A second solution of the multiple-access problem is to employ some form of *random-access*, i.e., to permit any sender to seize the entire communications resource when he happens to have information to transmit. The random-access solution is actually older than the multiplexing solution. For instance, the technique by which "ham" operators share a particular radio frequency channel is a random-access one. If two hams come up on the channel at virtually the same time, their transmissions interfere. But the inherent randomness in human actions ensures that eventually one will repeat his call well enough in advance of the other that the latter hears the former's signal and remains quiet, allowing the former to seize the channel. Moreover, essentially the same random-access technique is used by many people around the same table to communicate with one another over the same acoustical channel.

A better name for the time-division multiplexing (either with or without demand assignment) solution to the multi-access problem might be *scheduled-access*. Any sender knows that eventually he will be granted sole access to the channel, perhaps to send some information or perhaps to ask for a larger share of the resources. The key consequence is that the resources will be wasted during the period that he is granted sole access when in fact he has nothing to say. Thus, scheduled-access techniques tend to become inefficient when there are a large number of senders, each of which has nothing to say most of the time. But this is just the situation where random-access techniques tend to become efficient.

The computer age has given rise to many multiple-access situations in which there are a large number of senders, each of which has nothing to say most of the time. One such situation, namely the problem of communicating from remote terminals on various islands of Hawaii via a common radio channel to the main computer, led to the invention of Abramson [1] of the first formal random-access algorithm, now commonly called *pure Aloha*. Here it is supposed that transmitters can send data only in packets of some fixed duration, say T seconds. In pure Aloha, a transmitter always transmits such a packet at the moment it is presented to the transmitter by its associated information source. If no other transmitter is active during this T second transmission, then the packet is successfully received. Otherwise, there is a "collision" that is assumed to destroy all the packets that overlap. It is further supposed that, via some form of feedback, the transmitters discover whether or not their packets suffer collisions. When a collision occurs, the packets must be retransmitted. To avoid a repetition of the same collision, pure Aloha specifies that after a collision each transmitter involved randomly selects a waiting time before it again retransmits its packet. Assuming a Poisson traffic model and "statistical equilibrium", Abramson showed that pure Aloha had a maximum "throughput" of $1/2e \approx .184$, computed as the fraction of time on the channel occupied by successfully transmitted packets. It was soon noticed by Roberts [2], that the maximum throughput could be doubled to $1/e \approx .368$ by "slotting time" into T second intervals and requiring that the transmission of a packet be slightly delayed (if necessary) to coincide with a slot. This modification to Abramson's algorithm is now known as the *slotted-Aloha* random-access algorithm.

In his 1970 paper that first proposed the pure Aloha algorithm, Abramson introduced the hypothesis of "statistical equilibrium" in order to analyze the algorithm's performance for a Poisson traffic model. Essentially, this hypothesis states that the algorithm will eventually reach a steady-state situation in which the traffic from retransmitting of messages will form a stationary Poisson process that is independent of the new message traffic. It is precisely this assumption that leads to the maximum throughput bounds of 1/2e and 1/e for pure Aloha and slotted-Aloha, respectively. Abramson's statistical equilibrium assumption was a bold one and was best justified by the fact that, without it, the analytical tools appropriate for treating his algorithm did not exist. As time went on, however, communications engineers generally forgot that there was neither mathematical nor experimental justification for this hypothesis of "statistical equilibrium", and came to accept the numbers 1/2e and 1/e as the "capacities" of the pure Aloha channel and slotted Aloha channel, respectively, for the Poisson traffic model. Even more unfortunately, most workers continued to invoke the hypothesis of "statistical equilibrium" to "prove" that their particular rococo extension of the Aloha algorithm had superior delay-throughput properties compared to all previous ones, even though the character of their refinements should have made the hypothesis all the more suspect.

The next breath of truly fresh air in the research on random-access algorithms came in a 1977 M.I.T. doctoral dissertation by Capetanakis [3]. Capetanakis departed from the path beaten by Abramson in two important ways. First, he showed how, without prior scheduling or central control, the transmitters with packets to retransmit could make use of the known past history of collisions to cooperate in getting those packets through the channel. Second, he eliminated statistical equilibrium as an *hypothesis* by *proving* mathematically that his algorithm would reach a steady-state, albeit a highly non-Poisson one for the retransmitted traffic, when the new packet process was Poisson with a rate less than some specific limit. It must have come as a bombshell to many that Capetanakis could prove that his scheme achieves throughputs above the 1/e "barrier" for slotted-Aloha.

The aim of this paper is to illuminate the key features of Capetanakis' work and the subsequent work by others based on it, and to expose some analytical methods that appear useful in such studies. We also introduce a few new results of our own.

In Section 2 we formulate the concept of a "collision-resolution algorithm" and treat Capetanakis' algorithm within that context. In Section 3, we catalog those properties of the Capetanakis collision-resolution algorithm that are independent of the random process describing the generation of new packets. Then, in Section 4, we use these properties to analyze the performance of the Capetanakis' random-access algorithm (and its variants) for the Poisson traffic model under idealized conditions. In Section 5, we study quantitatively the effects of relaxing the idealized assumptions to admit propagation delays, channel errors, etc. Finally, we summarize the main conclusion of our study and review the historical development of the central concepts, hopefully with due credit to the original contributors.

2. COLLISION-RESOLUTION ALGORITHMS

2.1. General Assumptions

We wish to consider the random-accessing by many transmitters of a common receiver under the following idealized conditions:

(i) The forward channel to the receiver is a *time-slotted collison-type channel,* but is otherwise noiseless. The transmitters can transmit only in "packets" whose duration is one slot. A "collision" between two or more packets is always detected as such at the receiver, but the individual packets cannot be reconstructed at the receiver.

(ii) The *feedback channel* from the common receiver is a noiseless broadcast channel that informs the transmitters immediately at the end of each slot whether (a) that slot was empty, or (b) that slot contained one packet (which was thus successfully transmitted), or (c) that slot contained a collision of two or more packets (which must thus be retransmitted at later times.)

(iii) *Propagation delays are negligible,* so that the feedback information for slot i can be used to determine who should transmit in the following slot.

In later sections, we shall relax each of these conditions to obtain a more realistic model of a random access system. We shall see, however, that the analysis for the idealized case can be readily generalized to incorporate more realistic assumptions.

2.2. Definition of a Collision-Resolution Algorithm

By a collision-resolution algorithm for the random-accessing of a collision-type channel with feedback, we mean a protocol for the transmission and retransmission of packets by the individual transmitters with the property that after each collision all packets involved in the collision are eventually retransmitted successfully and all transmitters (not only those whose packets collided) eventually and simultaneously become aware that these packets have been successfully retransmitted. We will say that the collision is resolved precisely at the point where all the transmitters simultaneously become aware that the colliding packets have all been successfully retransmitted.

It is not at all obvious that collision-resolution algorithms exist. The Aloha algorithm, for instance, is not a collision-resolution algorithm as one can never be sure that all the packets involved in any collision have been successfully transmitted. Thus, the recent discovery by Capetanakis [3–5] of a collision-resolution algorithm was a surprising development in the evolution of random-access techniques whose full impact has yet to be felt.

It might seem that a collision-resolution algorithm would require a freeze after a collision on the transmission of new packets until the collision had been resolved. In fact, Capetanakis' algorithm does impose such a freeze. However, as we shall see later, one can devise collision-resolution algorithms that incorporate no freeze on new packet transmissions – another somewhat surprising fact.

2.3. The Capetanakis Collision-Resolution Algorithm (CCRA)

In [3,4], Capetanakis introduced the collision-resolution algorithm of central interest in this paper and which he called the "serial tree algorithm". We shall refer to this algorithm as the *Capetanakis collision-resolution algorithm* (CCRA). The CCRA can be stated as follows:

> **CCRA:** After a collision, all transmitters involved flip a binary fair coin; those flippping 0 retransmit in the very next slot, those flipping 1 retransmit in the next slot after the collision (if any) among those flipping 0 has been resolved. No new packets may be transmitted until after the initial collision has been resolved.

The following example should both clarify the algorithm and illustrate its main features.

Suppose that the initial collision is among 4 transmitters, as shown in Figure 2.1. For

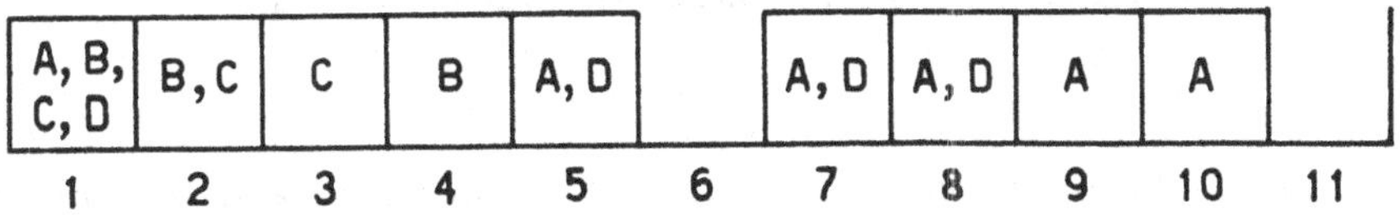

Fig. 2.1: Example of a Collision Resolution Interval for the CCRA.

convenience, we refer to these transmitters as A, B, C and D.

After the collision in slot 1, all four of these transmitters flip their coins – we suppose that B and C flip 0 while A and D flip 1. Thus B and C flip again at the end of slot 2 – we suppose that C flips 0 while B flips 1. Thus, only C sends in slot 3 and his packet is now successfully transmitted. B thus recognizes that he should send in slot 4, and his packet is now successfully transmitted.

It is illuminating to study the action thus far in the algorithm on the tree diagram in Figure 2.2 in which the number above each node indicates the time slot, the number inside the node indicates the feedback information for that time slot (0 = empty slot, 1 = single packet, $\geqslant 2$ = collision), and the binary numbers on the branches coming from a node indicate the path followed by these transmitters that flipped that binary number after the collision at that node. Thus, *collisions correspond to intermediate nodes* in this binary rooted tree since such nodes must be extended by the algorithm before the collision at that

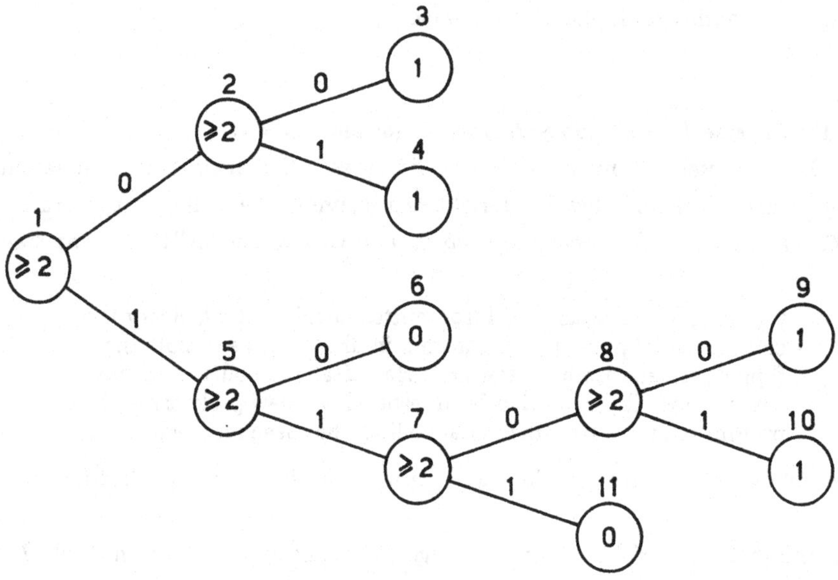

Fig. 2.2: Tree Diagram for the Collision Resolution Interval of Fig. 2.1.

node is resolved. On the other hand, *empty slots or slots with only one packet correspond to terminal nodes* in this binary rooted tree because after the corresponding transmission *all* transmitters simultaneously learn that any transmitter sending in that slot (i.e., zero or one transmitters) has successfully transmitted his message. Thus, a collision is resolved when and only when the algorithm has advanced to the point that the corresponding node forms the root node of a completed binary subtree. Thus, from Figure 2.2, which illustrates the same situation as Figure 2.1, we see that the collision in slot 2 is resolved in slot 4. Thus, transmitters A and D, who have been patiently waiting since the collision in slot 1, recognize that they should now retransmit in slot 5.

After the collision in slot 5, we suppose that A and D both flip 1. Thus, slot 6 is empty, so A and D again recognize that they should retransmit in slot 7. After the collision in slot 7, we suppose that A and D both flip 0. Hence they both retransmit in slot 8. After the collision in slot 8, we suppose that A flips 0 and D flips 1. Thus, A successfully transmits in slot 9 and D successfully transmits in slot 10. All four transmitters in the original collision have now transmitted successfully, but the collision is not yet resolved. The reason is that no one can be sure that there was not another transmitter, say E, who transmitted in slot 1, then flipped 1 and who retransmitted in slot 5 and then flipped 1 and who retransmitted again in slot 7 and then flipped 1 and who thus is now waiting to retransmit in slot 11. It is not until slot 11 proves to be empty that the original collision is finally resolved. All transmitters (not just the four in the original collision) can grow the tree in Figure 2.2 from the feedback information, and thus all transmitters now simultaneously learn that all the packets in the original collision have now been successfully transmitted.

Because a binary rooted tree (i.e., a tree in which either 2 or no branches extend from the root and from each subsequent node) has exactly one more terminal node than it has intermediate nodes (or two more terminal nodes then intermediate nodes excluding the root node), we have the following:

> **CCRA Property:** A collision in some slot is resolved precisely when the number of subsequent collision-free slots exceeds by two the number of subsequent slots with collisions.

For instance, from Figure 2.1 we see that the collision in slot 2 is resolved in slot 4, the collision in slot 5 is resolved in slot 11, the collision in slot 1 is also resolved in slot 11, etc. Notice that the later collisions are resolved sooner.

The above CCRA Property suggests a simple way to implement the CCRA due to R.G. Gallager [6].

> **CCRA Implementation:** When a transmitter flips 1 following a collision in which he is involved, he sets a counter to 1, then increments it by one for each subsequent collision slot and decrements it by one for each subsequent collision-free slot. When the counter reaches 0, the transmitter retransmits in the next slot.

Additionally, all transmitters must know when the original collision (if any) has been resolved as this determines when new packets may be sent. For this purpose, it suffices for each transmitter to have a *second counter* which is set to 1 just prior to the first slot, then incremented by one for each subsequent collision slot and decremented by one for each subsequent collision-free slot. When this second counter reaches 0, the original collision (if any) has been resolved.

We shall refer to the period beginning with the slot containing the original collision (if any) and ending with the slot in which the original collision is resolved (or ending with the first slot when that slot is collision-free) as the *collision-resolution interval* (CRI). Our main interest will be in the statistics of the *length* of the CRI, i.e., of the number of slots in the CRI. Note that the CRI illustrated by Figure 2.1 (or, equivalently, by Figure 2.2) has length 11 slots. Since 4 packets are successfully transmitted in this CRI, its "throughput" is $4/11 \approx .364$ packets/slot. We shall soon see that this is quite typical for the throughput of a CRI with 4 packets in the first slot when the CCRA is used to resolve the collision.

It should also be noted that, in the implementation of the CCRA, the feedback information is used only to determine whether the corresponding slot has a collision or is collision-free. *It is not necessary, therefore, to have the feedback information distinguish empty slots from slots with one packet when the CCRA is used.*

2.4. The Modified Capetanakis Collision-Resolution Algorithm (see also [7])

Referring to the example in Figure 2.2, we see that after slot 6 proves to be empty following the collision in slot 5, all transmitters now know that all the packets which collided in slot 5 will be retransmitted in slot 7. Thus, all transmitters know in advance that slot 7 will contain a collision. (Note that this statement will only be true when the feedback information distinguishes empty slots from slots with one packet.) Thus, it is wasteful actually to retransmit these packets in slot 7. The transmitters can "pretend" that this collision has taken place and immediately flip their binary coins and continue with the CCRA. The suggestion to eliminate these "certain-to-contain-a-collision" slots in the CCRA is due to the author. We shall refer to the corresponding algorithm as the modified Capetanakis collision-resolution algorithm (MCCRA). The MCCRA may be stated as follows:

> **MCCRA:** Same as the CCRA algorithm *except* that when the feedback indicates that a slot in which a set of transmitters who flipped 0 should retransmit is in fact empty, then each transmitter involved in the most recent collision flips a binary fair coin, those flipping 0 retransmit in the very next slot, those flipping 1 retransmit in the next slot after the collision (if any) among those flipping 0 is resolved (subject to the exception above.)

Figure 2.3 gives the binary tree for a CRI containing two packets in the original

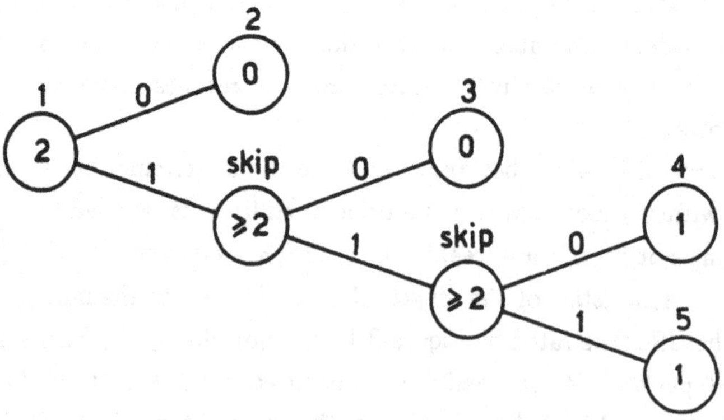

Fig. 2.3: Tree Diagram for a Collision Resolution Interval with the MCCRA.

collision and for which both transmitters flipped 1 on the first two tosses of their binary coin, but one flipped a 0 and the other a 1 on their third toss. The nodes labelled "skip" and having no slot number written above them correspond to points where the feedback indicates that certain transmitters should immediately flip their binary coins to thwart a certain collision. Note that this CRI has length 5, but would have had length 7 if the unmodified CCRA had been used because then the nodes labelled "skip" would become collision slots.

We observe next that, in the MCCRA, *an empty slot corresponding to retransmissions by a set of transmitters who flipped 0 is precisely the same as an empty slot that is separated from the most recent collision only by empty slots.* This follows from the facts that transmitters who flip 0 always send in the slot immediately after their flip is made, and that a flip is made only after a collision or after a "skipped collision", i.e., after an empty slot corresponding to retransmissions by a set of transmitters who had flipped 0. This observation justifies the following:

> **MCCRA Implementation:** Each transmitter has a flag F that is initially 0 and that he sets to 1 after a collision slot and sets to 0 after a slot with one packet. When a transmitter flips 1 following a collision in which he is involved, one for each subsequent collision, decrements it by 1 for each subsequent slot with one pocket, and also decrements it by one for each subsequent empty slot that occurs with F = 0. If his counter is 1 after an empty slot that occurs with F = 1 he flips his binary coin and decrements his counter by one if and only if 1 was flipped. When the counter reaches 0, the transmitter retransmits in the next slot.

Again the transmitters can use a *second counter* in conjunction with the same flag to determine when the CRI is complete. This second counter is set to 1 prior to the first slot, then incremented by one for each subsequent collision slot, decremented by one for each subsequent slot with one packet, and also decremented by one for each subsequent empty slot that occurs with F = 0. When this second counter reaches 0, the original cellision (if any) has been resolved.

Because the MCCRA is merely the CCRA modified to eliminate slots where collisions are certain to occur, it will always perform at least as well as the latter algorithm. In the MCCRA, we appear to have "gotten something for nothing". This is not quite true, however, for two minor reasons:

(i) The MCCRA, unlike the CCRA, requires the feedback information to distinguish between empty slots and slots with one packet.

(ii) The MCCRA is slightly more complex to implement than the CCRA because of the necessity for the "flag" in the former algorithm.

But there is a third and far stronger reason for hesitation in preferrring the MCCRA over the CCRA that we will subsequently demonstrate but that is not apparent under the idealized conditions considered in this section, namely:

(iii) When channel errors can occur, the MCCRA can suffer *deadlock*, i.e., reach a situation where the CRI never terminates and no packets are ever transmitted after some point.

3. TRAFFIC-INDEPENDENT PROPERTIES OF THE CAPETANAKIS COLLISION-RESOLUTION ALGORITHM

3.1. Definitions

In both the Capetanakis collision-resolution algorithm (CCRA) and its modification, when there is a collision in the first slot of a collision-resolution interval (CRI) no new packets may be transmitted until the CRI is completed. We shall let X denote the *number of packets transmitted in the first slot* of some CRI, and let Y denote the *length (in slots) of this same CRI.* Given X, Y depends only on the results of the coin tosses performed internally in the algorithms, and hence is independent of the traffic statistics that led to the given value of X. We can thus refer to any statistic of the CRI conditioned upon the value of X as *traffic-independent.* In this section, we shall study those traffic-independent properties of the CCRA and the MCCRA that are of greatest importance for the performance of random-access algorithms that incorporate these algorithms. The first and most important of these is the *conditional mean CRI length,* L_N, defined as

$$L_N = E(Y|X = N). \tag{3.1}$$

The *conditional second moment of the CRI length,* S_N, defined by

$$S_N = E(Y^2|X = N), \tag{3.2}$$

is also of fundamental importance. The *conditional variance of CRI length,* V_N, defined by

$$V_N = Var(Y|X = N), \tag{3.3}$$

will prove to be more tractable than S_N directly, but of course is related to S_N by

$$V_N = S_N - (L_N)^2. \tag{3.4}$$

We now investigate these traffic-independent quantities in detail for the CCRA.

3.2 Intuitive Analysis

Our aim here is to determine the coarse dependence of L_N, V_N and S_N on N for the CCRA as a guide to a subsequent precise analysis. To do this, we suppose that $N = 2n$ is very large. Then, as shown in Figure 3.1, there will be a collision in slot 1 of the CRI, following which very close to half of the transmitters will flip 0 and half will flip 1. Thus

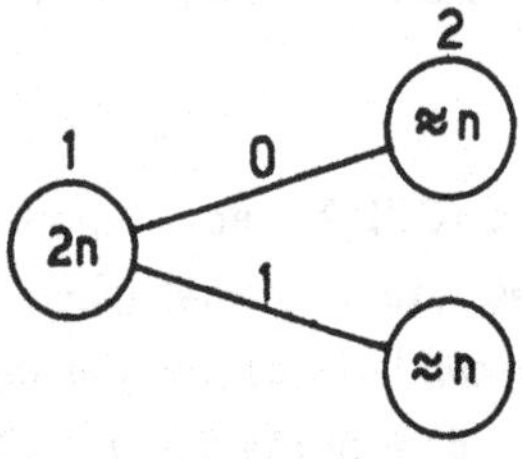

Fig. 3.1: Typical Action of the CCRA in First Slot of the CRI When n Is Large.

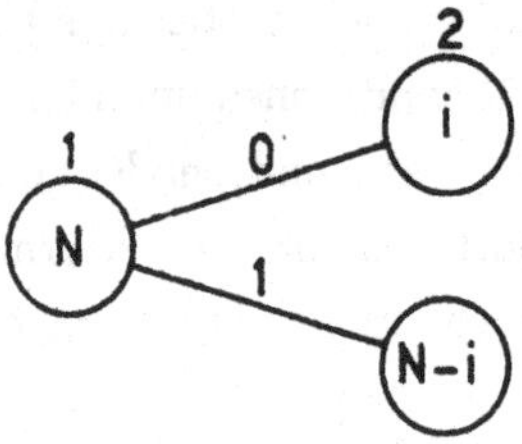

Fig. 3.2: General Action of the CCRA in First Slot of CRI when N Exceeds 1.

(3.5) $$L_{2n} \approx 1 + 2L_n, \quad n >> 1$$

since the expected number of slots needed to resolve the collision in slot 2 of the approximately n transmitters who flipped 0 is L_n, following which the expected number of slots to resolve the subsequent collision of the approximately n transmitters who flipped 1 is also L_n. Considered as an equality, (3.5) is a recursion whose solution is $L_N = \alpha N - 1$ for an arbitrary constant . Thus, we conclude that

(3.6) $$L_n \approx \alpha N - 1, \quad N >> 1$$

describes the coarse dependence of L_N on N, which whets our appetite to find the constant α. In fact, we will soon see that (3.6) is remarkably accurate even for small values of N.

From Figure 2.1 and the fact that the number of slots needed to resolve the collision among the n transmitters who flipped 0 is independent of the number of slots needed to resolve the collision among the n transmitters who flipped 1, we see that

(3.7) $$V_{2n} \approx 2V_n, \quad n >> 1.$$

This recursion forces the conclusion that

(3.8) $$V_N \approx \beta N, \quad N >> 1$$

for some constant β. We shall soon see that (3.8) likewise is quite accurate even for rather small N. Finally, (3.4), (3.6) and (3.8) imply

(3.9) $$S_N \approx \alpha^2 N^2 - (2\alpha - \beta)N + 1, \quad N >> 1,$$

which completes our intuitive analysis.

3.3. Conditional Mean of CRI Length

We now give for the CCRA a precise analysis of the expected CRI length, L_N, given that X = N packets are transmitted in the first slot. When N is 0 or 1, the CRI ends with its first slot so that

$$L_0 = L_1 = 1. \tag{3.10}$$

When $N \geq 2$, there is a collision in the first slot. The probability that exactly i of the colliding transmitters flip 0 (as depicted in Figure 3.2) is just

$$p_N^{(i)} = \binom{N}{i} 2^{-N}, \tag{3.11}$$

in which case the expected CRI length is just $1 + L_i + L_{N-i}$. Hence, it follows that

$$\begin{aligned} L_N &= 1 + \sum_{i=0}^{N} (L_i + L_{N-i})\, p_N^{(i)} \\ &= 1 + 2 \sum_{i=0}^{N} L_i p_N^{(i)} \end{aligned} \tag{3.12}$$

where we have used the fact that $p_N(i) = p_N(N-i)$. Solving for L_N gives the recursion

$$L_N = [1 + 2 \sum_{i=0}^{N-1} L_i p_N^{(i)}]/(1 - 2^{-N+1}) \tag{3.13}$$

which holds for all $N \geq 2$. The required initial conditions are given in (3.10). In Table 3.1, we give the first vew values of L_N as found from (3.13).

N	L_N
0	1
1	1
2	5
3	$23/3 \approx 7.667$
4	$221/21 \approx 10.524$
5	13.419
6	16.313

Table 3.1 Expected CRI length for the CCRA given N packets in the first slot.

From Table 3.1, we can see that $L_N - L_{N-1} \approx 2.9$ for $N \geqslant 4$. This suggests that the constant α in (3.6) is about 2.9. In fact, we see from Table 3.1 that the $N>>1$ approximation $L_N \approx 2.9\,\alpha - 1$ is already quite accurate for $N \geqslant 3$.

We now develop a technique for obtaining arbitrarily tight upper and lower bounds on L_N. We begin by choosing an arbitrary positive integer M. We seek to find a constant, α_{uM}, as small as possible, for which we can prove

$$L_N \leqslant \alpha_{uM} N - 1, \quad \text{all } N \geqslant M. \tag{3.14}$$

(The first subscript on α is only to remind us that this constant appears in an *upper* bound on L_N.) By using the Kronecker delta δ_{ij} defined to be 1 if $i = j$ and 0 if $i \neq j$, we can rewrite (3.14) as

$$L_N \leqslant \alpha_{uM} N - 1 + \sum_{i=0}^{M-1} \delta_{iN} (L_N - \alpha_{uM} N + 1), \quad \text{all } N \tag{3.15}$$

because the right side reduces to L_N for $N < M$. Substituting the bound (3.15) for those L_i on the right in (3.13) and making use of the fact that

$$\sum_{i=0}^{N-1} i p_N(i) = \sum_{i=0}^{N} i p_N(i) - N p_N(N)$$

$$= \frac{N}{2}(1 - 2^{-N+1}),$$

we obtain

$$L_N \leqslant \alpha_{uM} N - 1 + 2\left[\sum_{i=0}^{M-1} (L_i - \alpha_{uM} i + 1) p_N(i)\right] / (1 - 2^{-N+1}). \tag{3.16}$$

It thus follows by induction that (3.14) holds for any α_{uM} such that the summation in square brackets on the right in (3.16) is nonpositive for all $N \geqslant M$, i.e., such that

$$\alpha_{uM} \sum_{i=0}^{M-1} i p_N(i) \geqslant \sum_{i=0}^{M-1} (L_i + 1) p_N(i), \quad N \geqslant M. \tag{3.17}$$

The best upper bound is obtained by choosing α_{uM} such that (3.17) holds with equality in the "worst case", i.e., choosing

$$\alpha_{uM} = \sup_{N \geqslant M} \left[\sum_{i=0}^{M-1} \binom{N}{i} (L_i + 1) / \sum_{i=0}^{M-1} \binom{N}{i} \, i \right]. \tag{3.18}$$

By an entirely analogous argument, one can show inductively that

$$L_N \geqslant \alpha_{\ell M} N - 1, \quad \text{all } N \geqslant M$$

holds for the choice

$$\alpha_{\ell M} = \inf_{N \geqslant M} \left[\sum_{i=0}^{M-1} \binom{N}{i} (L_i + 1) / \sum_{i=0}^{M-1} \binom{N}{i} i \right] . \tag{3.19}$$

For a given M, after one has calculated L_i for all $i < M$, it is a simple matter numerically to find the maximizing N and minimizing N in (3.18) and (3.19), respectively, and hence to determine α_{uM} and $\alpha_{\ell M}$. Table 3.2 summarizes these calculations and

M	α_{iM}	Maximizing N in (3.18)	$\alpha_{\ell M}$	Minimizing N in (3.19)
2	3	2	2	∞
3	3	∞	2.8750	4
4	2.8965	14	2.8810	4
5	2.8867	8	2.8810	∞

Table 3.2 Values of the coefficients α_{uM} and $\alpha_{\ell M}$ in the bounds (3.18) and (3.19), respectively, for L_N.

furnishes bound on the "true coefficient" α in (3.6). From the M = 5 case, we see that

$$2.8810 \leqslant \alpha \leqslant 2.8867 \tag{3.20}$$

so that in fact we know the value of α to three significant decimal digits. We can also summarize the M = 5 results as

$$2.8810N - 1 \leqslant L_N \leqslant 2.8867N - 1, \quad N \geqslant 4. \tag{3.21}$$

(The inductive argument for M = 5 guarantees that (3.21) holds for $N \geqslant 5$; by checking against the values in Table 3.1, we find that (3.21) holds also for N = 4.) For all practical purposes, the bounds in (3.21) are so tight as to be tantamount to equalities, and we conclude that we now have determined L_N for the CCRA and are ready to move on to consider the conditional second moment of epoch length. Before doing so, it is interesting to note (as was pointed out to us by W. Sandrin of the Comsat Laboratories) that

$$\frac{2}{\ln 2} = 2.8854 \tag{3.22}$$

which, together with the binary nature of the CCRA, lends support to the conjecture that

$$\alpha = \lim_{M\to\infty} \alpha_{uM} = \lim_{M\to\infty} \alpha_{\ell M} = \frac{2}{\ln 2}\,. \tag{3.23}$$

3.4. Conditional Variance and Second Moment of CRI Length

We now seek tight bounds on the conditional variance, $V_N = \mathrm{Var}(Y \mid X = N)$, and the conditional second moment, $S_N = E(Y^2 \mid X = N)$, of the CRI length for the CCRA. Letting X_0 denote the number of transmitters who flip 0 after the collision in slot 1, we see from Figure 3.2 that

$$\begin{aligned} S_N &= \sum_{i=0}^{N} E(Y^2 \mid X = N, X_0 = i)p_N(i) \\ &= \sum_{i=0}^{N} [\mathrm{Var}(Y \mid X = N, X_0 = i) + E^2(Y \mid X = N, X_0 = i)]p_N(i) \\ &= \sum_{i=0}^{N} [V_i + V_{N-i} + (1 + L_i + L_{N-i})^2]p_N(i) \\ &= \sum_{i=0}^{N} [2V_i + (1 + L_i + L_{N-i})^2]p_N(i) \end{aligned} \tag{3.24}$$

for $N \geqslant 2$, where the next-to-last equality follows from the fact that the number of slots used to resolve the collision (if any) among the i transmitters who flipped 0 is independent of the number used to resolve the collision (if any) among the $N - i$ transmitters who flipped 1, and the last equality follows from the fact that $p_N(i) = p_N(N - i)$. Combining (3.4) and (3.24), we obtain

$$V_N = [2\sum_{i=0}^{N-1} V_i p_N(i) + \sum_{i=0}^{N} (1 + L_i + L_{N-i})^2 p_N(i) - L_N^2]/(1-2^{-N+1}) \tag{3.25}$$

for $N \geqslant 2$, which is our desired recursion for V_N. Because $Y = 1$ when $X = 0$ or $X = 1$, the appropriate initial conditions are

$$V_0 = V_1 = 0\,. \tag{3.26}$$

Using the values of L_N given in Table 3.1, we can use (3.25) to find the values of V_N

given in Table 3.3, to which we have added the corresponding values of S_N found using

N	V_N	S_N
0	0	1
1	0	1
2	8	33
3	$88/9 \approx 9.78$	68.56
4	13.53	124.2
5	16.93	197.0
6	20.32	286.3

Table 3.3 Variance and second moment of CRI length for the CCRA given N packets in the first slot.

(3.4). From Table 3.3, we see that the linear growth of V_N with N predicted asymptotically by (3.8) is already evident for $N \geqslant 4$, the constant of proportionality being $\beta \approx 3.4$.

We can develop a simple lower bound on V_N by noting that, because x^2 is a convex function, Jensen's inequality [8] together with (3.12) implies

$$\sum_{i=0}^{N} (1 + L_i + L_{N-i})^2 \, p_N(i) \geqslant L_N^2 \, . \tag{3.27}$$

Substituting (3.27) in (3.25) gives the simple inequality

$$V_N \geqslant 2 \sum_{i=0}^{N-1} V_i p_N(i) \, / \, (1 - 2^{-N+1}) \, . \tag{3.28}$$

In the same manner as led to (3.19), we can use (3.28) to verify that

$$V_N \geqslant \beta_{\ell M} N \, , \qquad N \geqslant M \tag{3.29}$$

holds for the choice

$$\beta_{\ell M} = \inf_{N \geqslant M} \left[\sum_{i=0}^{M-1} \binom{N}{i} V_i \sum_{i=0}^{M-1} \binom{N}{i} i \right] . \tag{3.30}$$

Table 3.4 gives the values of $\beta_{\ell M}$ for $3 \leqslant M \leqslant 6$.

M	$\beta_{\ell M}$	Minimizing N in (3.30)	β_{uM}	Maximizing N in (3.36)
3	2.666	3	4	∞
4	3.111	4	3.506	4
5	3.272	5	3.458	5
6	3.333	6	3.424	6
7	3.359	7	3.404	7

Table 3.4 Values of the coefficients $\beta_{\ell M}$ and β_{uM} in the bounds (3.30) and (3.36) respectively, on V_N.

To obtain an upper bound on V_N, we first observe that

$$L_i + L_{N-i} < L_N \,, \qquad 0 < i < N. \tag{3.31}$$

A proof of (3.31) is unwardingly tedious and will be omitted, but its obviousness can be seen from the fact that it merely states that the CCRA will do better in processing two non-empty sets of transmitters in separate CRI's than it would do if the two sets first were merged and then the CCRA applied. In fact, it should be obvious from the tight bounds on L_N developed in the previous section that the right side of (3.31) will exceed the left by very close to unity as soon as both i and N − i are four or greater, and this can be made the basis of a rigorous proof of (3.31). Next, we observe that (3.31) when $i = 1$ or $i = N - 1$ can be strengthened to

$$L_1 + L_{N-1} < L_N - 1 \,, \tag{3.32}$$

as follows from the fact that $L_1 = 1$ so that the right side of 3.32 will exceed the left by approximately $\alpha - 2 \approx 0.9$ when N is four or greater – the validity of (3.32) for $N < 4$ can be checked from Table 3.1.

Using (3.31), (3.32) and the fact that $p_N(1) = Np_N(0) \geqslant 3p_N(0)$ for $N \geqslant 3$, one easily finds

$$\sum_{i=0}^{N} (1 + L_i + L_{N-i})^2 p_N(i) < (L_N + 1)^2 \,, \qquad N \geqslant 3. \tag{3.33}$$

Substituting (3.33) into (3.25) gives

$$V_N < \left[2 \sum_{i=0}^{N-1} V_i p_N(i) + 2L_N + 1 \right] / (1 - 2^{-N+1}) \,, \qquad N \geqslant 3 \tag{3.34}$$

which is our desired simple upper bound on V_N. In the now familiar manner we can use (3.34) to verify that

$$V_N \leqslant \beta_{uM} N, \qquad N \geqslant M \tag{3.35}$$

holds for the choice

$$\beta_{uM} = \sup_{N \geqslant M} \left\{ \left[\sum_{i=0}^{M-1} \binom{N}{i} \quad V_i + \alpha_M N - \frac{1}{2} \right] \Big/ \left[\sum_{i=0}^{M-1} \binom{N}{i} \quad i \right] \right\} \tag{3.36}$$

for $M \geqslant 3$, where here α_M is any constant for which it is known that $L_N \leqslant \alpha_M N - 1$ for $N \geqslant M$. Taking $\alpha = 2.90$ for $M = 3$ and $\alpha = 2.8867$ for $M \geqslant 4$ (as were justified in the previous section), (3.36) results in the values of β_{uM} as given in Table 3.4.

The $M = 7$ cases in Table 3.4 provide the bounds

$$3.359N \leqslant V_N \leqslant 3.404N, \qquad N \geqslant 4. \tag{3.37}$$

The validity of (3.36) for $N < M$, i.e., for $N = 4,5$ and 6 may be directly checked from Table 3.3. Although (3.37) is not quite so tight as our corresponding bounds in (3.21) on L_N, it is tight enough to confirm our earlier suspicion that $\beta \approx 3.4$ and more than tight enough for our later computations. From (3.21), (3.37) and (3.4), we find the corresponding bounds on S_N to be

$$8.300N^2 - 2.403N + 1 \leqslant S_N \leqslant 8.333N^2 - 2.369N + 1, \qquad N \geqslant 3. \tag{3.38}$$

which, since the coefficient of N^2 is $\alpha_{\ell M}^2$ and α_{uM}^2 on the left and right sides, respectively, shows that the tightness of these bounds on S_N depends much more on the tightness of (3.21) than on that of (3.37).

3.5. Conditional Distribution of CRI Length

In this section, we consider the probability distributions, $P_{Y|X}(\ell|N)$, for the CRI length Y given that $X = N$ transmitters collide in the first slot and the CCRA is used. The cases $X = 0$ and $X = 1$ are trivial and give

$$P_{Y|X}(1|0) = P_{Y|X}(1|1) = 1. \tag{3.39}$$

For $N \geqslant 2$, however, every sufficiently large *odd* integer is a possible value of i. That Y must be odd follows from the facts that a binary rooted tree always has an odd number of nodes and that the slots in a CRI for the CCRA corresponds to the nodes in such a tree.

Writing $P(2m+1|N)$ for brevity in place of $P_{Y|X}(2m+1|N)$, we note first that

$$P(3|2) = 1/2, \tag{3.40}$$

because with probability 1/2 the two transmitters colliding in slot 1 will flip different values and the CRI length is then 3. When these two transmitters flip the same value, then a blank slot together with another collision occurs and hence

$$P(2m+1|2) = \frac{1}{2} P(2(m-1)+1|2), \quad m \geqslant 2. \tag{3.41}$$

Equation (3.41) is a first-order linear recursion whose solution for the initial condition (3.40) is

$$P(2m+1|2) = 2^{-m}, \quad m \geqslant 1, \tag{3.42}$$

which shows that $(Y-1)/2$ is geometrically distributed when $X = 2$.

When $X = 3$, the minimum value of Y is 5 and occurs when and only when the three transmitters do not all flip the same value after the initial collision and the two who flip the same value then flip different values after their subsequent collision. This reasoning gives

$$P(5|3) = \frac{3}{4}\frac{1}{2} = \frac{3}{8}. \tag{3.43}$$

In general, we see that

$$P(2m+1|3) = \frac{1}{4} P(2(m-1)+1|3) + \frac{3}{4} P(2(m-1)|2)$$

for $m \geqslant 3$. With the aid of (3.41) this can be rewritten

$$P(2m+1|3) = \frac{1}{4} P(2(m-1)+1|3) = 3(2^{-m-1}), \quad m \geqslant 3. \tag{3.44}$$

Equation (3.43) is another first-order linear recursion whose solution for the initial condition (3.43) is

$$P(2m+1|3) = 3(2^{-m}) - 6(4^{-m}), \quad m \geqslant 2. \tag{3.45}$$

Continuing this same approach, one could find $P(2m + 1 | N)$ for all N. However, the number of terms in the resulting expression doubles with each increase of N so that the calculation fast becomes unrewardingly tedious. The above distributions for $N \leqslant 3$, however, are sufficiently simple in form that we can and will make convenient use of them in what follows.

This completes our analysis of the traffic-independent properties of the CCRA, and we now turn our attention to the MCCRA.

3.6. Corresponding Properties for the Modified CCRA

We now briefly treat the traffic-independent properties of the modified CCRA (MCCRA) that was introduced in Section 2.4. Using the same notation as was used for the CCRA, we first note that (3.12) for the MCCRA becomes

$$L_N = 1 + 2 \sum_{i=0}^{N} L_i p_N(i) - p_N(0) \tag{3.46}$$

for $N \geqslant 2$, as follows again from Figure 3.2 and the fact that when none of the transmitters in the initial collision flip 0 [which occurs with probability $p_N(0)$] then only $L_N - 1$ [rather than L_N] slots on the average are required to resolve the "collision" that was certain. Solving (3.46) for L_N gives the recursion

$$L_N = \left[1 + 2 \sum_{i=0}^{N-1} L_i p_N(i) - p_N(0) \right] / \, (1 - 2^{-N+1}) \tag{3.47}$$

which holds for all $N \geqslant 2$. The initial conditions are again

$$L_0 = L_1 = 1. \tag{3.48}$$

In Table 3.5, we give the first few values of L_N as found from (3.47). Comparing Tables 3.1 and 3.5, we see that the main effect of the slots saved by eliminating collisions in the MCCRA is to reduce L_2 from 5 to 9/2, and that this 10% savings propagates only slightly diminished to L_N with $N > 2$. We also note from Table 3.5 that $L_N - L_{N-1}$ 2.67 for $N \geqslant 4$.

Using precisely the same techniques as in Section 3.3, one easily finds for the MCCRA that

$$\alpha_{\ell M} N - 1 \leqslant L_N \leqslant \alpha_{uM} N - 1 , \quad N \geqslant M \tag{3.49}$$

N	N_L
0	1
1	1
2	9/2
3	7
4	9.643
5	12.314
6	14.985

Table 3.5 Expected CRI length for the MCCRA given N packets in the first slot

where $\alpha_{\ell M}$ and α_{uM} are the infimum and supremum, respectively, for $N \geqslant M$ of the function

$$f_M(N) = \left[\sum_{i=0}^{M-1} \binom{N}{i} (L_i + 1) - \frac{1}{2}\right] \Big/ \sum_{i=0}^{M-1} \binom{N}{i}\, i \,. \tag{3.50}$$

For M = 5, these bounds become

$$2.6607N - 1 \leqslant L_N \leqslant 2.6651N - 1\,, \quad N \geqslant 4 \tag{3.51}$$

where the fact that the bounds also hold for N = M − 1 = 4 can be checked directly from Table 3.5.

Equation (3.24), which gives S_N for the CCRA, is easily converted to apply to the MCCRA by noting that the only required change for the MCCRA is that $E^2(Y|X = N, X_0 = 0) = (L_0 + L_N)^2$ rather than $(1 + L_0 + L_N)^2$ because of the eliminated slot, so that (3.24) is changed to

$$S_N = \sum_{i=0}^{N} [2V_i + (1 + L_i + L_{N-i})^2] p_N(i) - (2L_N + 3)2^{-N} . \tag{3.52}$$

Beginning from (3.52), one can readily find the recursion for V_N analogous to (3.25), and then derive linear upper and lower bounds on V_N as was done in Section 3.4 for the CCRA. We shall, however, rest content with the bounds (3.51) on L_N, both because these are more important than those for V_N and also because our primary interest is in the CCRA rather than the MCCRA for the reason stated above at the end of Section 2.4.

4. RANDOM-ACCESS VIA COLLISION-RESOLUTION

4.1. The Obvious Random-Access Algorithm

We now consider the use of a collision-resolution algorithm as the key part of a random-access algorithm for the idealized situation described in Section 2.1. The principle is simple. To obtain a random-access algorithm from a collision-resolution algorithm, one needs only to specify the rule by which a transmitter with a new packet to send will determine the slot for its initial transmission. Thereafter, the transmitter uses the collision-resolution algorithm (if necessary) to determine when the packet should be retransmitted. One such first-time transmission rule is the obvious one: *transmit a new packet in the first slot following the collision-resolution interval* (CRI) *in progess when it arrived.* (Here we tacitly assume that no more than one new packet arrives at any transmitter while a CRI is in progress.) We shall refer to the random-access algorithm obtained in this way as the *obvious random-access algorithm* (ORAA) for the incorporated collision-resolution algorithm. That the ORAA may not be the ideal way to incorporate a collision-resolution algorithm is suggested by the fact that, after a very long CRI, a large number of packets will usually be transmitted in the next slot so that a collision there is virtually certain. Nonetheless, the ORAA is simple in concept and implementation, and is so natural that one would expect its analysis to yield useful insights.

For brevity, we shall write CORAA and MCORAA to denote the ORAA's incorporating the Capetanakis collision-resolution algorithm (CCRA) and the modified CCRA, respectively. We shall analyze the CORAA and the MCORAA in some detail before considering less obvious ways to incorporate collision-resolution algorithms in random-access algorithms.

4.2. Intuitive Stability Analysis

We assume that the random-access system is activated at time $t = 0$ with no backlog of traffic. The unit of time will be taken as one slot so that the i-th slot is the time interval $(i, i+1]$, $i = 0,1,2,\ldots$. The new packet process is a counting process N_t giving the number of new packets that arrive at their transmitters in the interval $(0,t]$. Thus, $N_{t+T} - N_t$ is the number of new packets that arrive in the interval $(t,t+T]$. We assume that the new packet process is characterized by a constant λ, the *new packet rate,* such that, for all $t \geqslant 0$, $(N_{t+T} - N_t)/T$ will be close to λ with high probability when λT is large. The action of the random-access algorithm on the arrival process N_t generates another counting process W_t giving the number of packets that have arrived at their transmitters in the interval $(0,t]$ but have not been successfully transmitted in this interval. The random-access algorithm is said to be *stable* or *unstable* according as whether W_t remains bounded or grows without bound with high probability as t increases.

The stability properties of the CORAA can be ascertained from the following intuitive argument. From (3.21) and (3.37), we see that both the mean L_N and variance V_N of CRI length for the CCRA grow linearly with the number, N, of packets ithe first slot. Thus, by Tchebycheff's inequality, the CRI length will be close to its mean $L_N \approx 2.89N$ with high probability when N is large. Now suppose some CRI begins in slot i and that W_i is large. By the first-time transmission rule for the OCRAA, all W_i packets will be transmitted in the first slot of the CRI so that the CRI length Y will be close to 2.89 W slots with high probability. But close to $\lambda Y = 2.89\lambda\ W_i$ new packets will with high probability arrive during the CRI so that

$$W_{i+Y} \approx 2.89\lambda\ W_i \tag{4.1}$$

with high probability. Thus, W_t will remain bounded with high probability, i.e., the CORAA will be stable, if

$$\lambda < \frac{1}{2.89} = .346 \text{ packets/slot}.$$

Conversely, provided only that the arrival process has "sufficient variability" so that W_i can in fact be large, (4.2) shows that the CORAA will be unstable for

$$\lambda > \frac{1}{2.89} = .346 \text{ packets/slot}.$$

The "sufficient variability" condition excludes deterministic arrival processes such as that with $N_i = i$ for $i = 1,2,3,\ldots$ which has $\lambda = 1$ and for which the CORAA is trivally stable. Other than for this restriction, the condition $\lambda < .346$ packet/slot is both necessary and sufficient for stability of the CORAA with very weak conditions on the arrival process N_t that we shall not attempt to make precise.

The same argument for the MCORAA, making use of (3.51) shows that

$$\lambda < \frac{1}{2.67} = .375 \text{ packets/slot}$$

is the corresponding sufficient condition for stability, and also a necessary condition when the arrival process has "sufficient variability". It is worth noting here that the upper limit of stability of .375 packets/slot for the MCORAA exceeds the $1/e = .368$ packets/slot "maximum throughput" of the unstable slotted-Aloha random-access algorithm.

In the following sections, we shall make the above stability arguments precise for the important case where the new packet process if Poisson. Along the way, we will determine the "delay-throughput characteristic" for the CORAA.

4.3. Dynamic Analysis of the CORAA

We now consider the case where the new packet process is a stationary Poisson point process, i.e., where $N_{t+T} - N_t$ is a Poisson random variable with mean λT for all positive T and all $t \geq 0$. Let Y_i and X_i denote the length and number of packets in the first slot, respectively, of the i-th CRI when the CORAA is applied, where Y_0 and X_0 correspond to the CRI beginning at $t = 0$. By assumption there are no packets awaiting transmission at $t = 0$ so that

$$X_0 = 0 \tag{4.2a}$$

$$Y_0 = 1. \tag{4.2b}$$

By the Poisson assumption, given that $Y_i = L$, X_{i+1} is a Poisson random variable with mean λL, i.e.,

$$P(X_{i+1} = N \mid Y_i = L) = \frac{(\lambda L)^N}{N!} e^{-\lambda L} \tag{4.3}$$

for $N = 0,1,2,\ldots$.

Because of (4.3) (which reflects the independent increments property of the Poisson new packet process), Y_{i+1} is independent of $Y_0, Y_1, \ldots Y_{i-1}$ when Y_i is given. Thus $Y_0, Y_1, Y_2, \ldots$ is a Markov chain, as is also the sequence $X_0, X_1, X_2, \ldots$. We now consider the "dynamic behavior" of these chains in the sense of the dependence of $E(X_i)$ and $E(Y_i)$ on i. We first note that the Poisson assumption implies

$$E(X_{i+1} \mid Y_i = L) = \lambda L, \tag{4.4}$$

which upon multiplication by $P(Y_i = L)$ and summing over L yields

$$E(X_{i+1}) = \lambda E(Y_i). \tag{4.5}$$

Equation (4.5) shows that finding the dependence of either $E(X_i)$ or $E(Y_i)$ on i determines the dependence of the other, so we will focus our attention on $E(X_i)$.

To illustrate our approach, we begin with the rather crude upper bound

$$L_N \leq 3N - 1 + 2\delta_{0N} - \delta_{1N}, \quad \text{all } N \geq 0 \tag{4.6}$$

for the CCRA, which follows from (3.18) with $M = 2$. But $E(Y_i|X_i = N) = L_N$ so that (4.6) implies

(4.7) $$E(Y_i | X_i = N) \leqslant 3N - 1 + 2\delta_{0N} - \delta_{1N}$$

Multiplying by $P(X_i = N)$ and summing over N gives

(4.8) $$E(Y_i) \leqslant 3\,E(X_i) - 1 + 2P(X_i = 0) - P(X_i = 1)\,.$$

Now using (4.5) in (4.8), overbounding $P(X_i = 0)$ by 1 and underbounding $P(X_i = 1)$ by 0, we obtain the recursive inequality

(4.9) $$E(X_{i+1}) - 3\lambda\,E(X_i) \leqslant \lambda\,.$$

When equality is taken in (4.9), we have a first-order linear recursion whose solution for the initial condition (4.2a) is an upper bound on $E(X_i)$, namely

(4.10) $$E(X_i) \leqslant \frac{\lambda}{1-3\lambda}\,[1-(3\lambda)^i], \qquad \text{all } i \geqslant 0$$

showing that $E(X_i)$ approaches a finite limit as $i \to \infty$ provided $\lambda < 1/3$. It is more interesting, however, to consider a "shift in the time origin" to allow

(4.11) $$X_0 = N$$

to be an arbitrary initial condition. The solution of (4.9) then yields the bound

(4.12) $$E(X_i) \leqslant N(3\lambda)^i + \frac{\lambda}{1-3\lambda}\,[1-(3\lambda)^i]\,, \qquad \text{all } i \geqslant 0\,.$$

Inequality (4.12) shows that when $\lambda < 1/3$ and X_0 is very large (as when we take the time origin to be at a point where momentarily a large number of packets are awaiting their first transmission), $E(X_i)$ approaches its asymptotic value of less than $\lambda/(1-3\lambda)$ exponentially fast in the CRI index i, and thus at least this fast in time t as the successive CRI's are decreasing in length on the average.

A similar argument beginning from the correspondingly crude lower bound for $M = 2$ in (3.18), namely

(4.13) $$L_N \geqslant 2N - 1 + 2\delta_{0N}$$

would give, for the initial condition (4.11), the bound

$$E(X_i) \geq N(2\lambda)^i - \frac{\lambda}{1-2\lambda}\,[1-(2\lambda)^i]\,, \qquad \text{all } i \geq 0, \tag{4.14}$$

showing that $E(X_i) \rightarrow \infty$ as $i \rightarrow \infty$ when $\lambda > 1/2$, and also showing for $\lambda < 1/2$ that the approach of $E(X_i)$ to its asymptotic value is not faster than exponential in the CRI index i.

It should be clear that had bounds (4.6) and (4.13) been replaced by the correspondingly sharp bounds from (3.21), we would have found that

$$E(X_\infty) \triangleq \lim_{i\rightarrow\infty} E(X_i) \tag{4.15}$$

is finite for

$$\lambda < \frac{1}{2.8867} = .3465 \qquad \text{(CORAA)} \tag{4.16}$$

but is infinite for

$$\lambda > \frac{1}{2.8810} = .3471 \qquad \text{(CORAA)} \tag{4.17}$$

Moreover, (4.5) implies that

$$E(Y_\infty) \triangleq \lim_{i\rightarrow\infty} E(Y_i) = \frac{1}{\lambda}\, E(X_\infty) \tag{4.18}$$

so that (4.16) and (4.17) are also conditions for the finiteness and nonfiniteness, respectively, of $E(Y_\infty)$.

Similar arguments based on (3.51) would have shown for the MCORRA that

$$\lambda < \frac{1}{2.6651} = .3752 \qquad \text{(MCORAA)} \tag{4.19}$$

and

$$\lambda > \frac{1}{2.6607} = .3758 \qquad \text{(MCORAA)} \tag{4.20}$$

imply the finiteness and non-finiteness, respectively, of $E(Y_\infty)$.

We will shortly see that conditions (4.16) and (4.17) in fact imply the stability and instability, respectively, of the CORAA corroborating the intuitive analysis of Section 4.2. Conditions (4.19) and (4.20) similarly imply the stability and instability, respectively, of the

MCORAA.

4.4. Stability Analysis of the CORAA

We have just seen that the Markov chain $X_0, X_1, X_2, \ldots$ [giving the number of packets in the first slots of the CRI's when the new packet traffic is Poisson and the CORAA is used] has $E(Y_\infty) < \infty$ when $\lambda < .3465$. We also note from (4.3) that, regardless of the value of X_i, X_{i+1} has nonzero probability of being any nonnegative integer. These two facts imply that for $\lambda < .3465$ the chain has *steady-state* probabilities

(4.21) $$\pi_N = P(X_\infty = N) \triangleq \lim_{i\to\infty} P(X_i = N) \qquad N = 0,1,2,\ldots$$

and is *ergodic* in the sense that if n_N is the number of CRI's with N packets among the first n CRI's, then

(4.22) $$\lim_{n\to\infty} \frac{n_N}{n} = \pi_N \qquad \text{(a.s.)}$$

Similarly, $\lambda < .3465$ implies that the Markov chain $Y_0, Y_1, Y_2, \ldots$ has steady-state probabilities

(4.23) $$P(Y_\infty = L) \triangleq \lim_{i\to\infty} P(Y_i = L) \qquad L = 1,3,5,\ldots$$

such that

(4.24) $$\lim_{n\to\infty} \frac{n'_L}{n} = P(Y_\infty = L)$$

where n_L is the number of CRI's of length L among the first n CRI's.

Let the random variable Y_a denote the length of the CRI in progress when a "randomly-chosen packet" *arrives* at its transmitter. Because the new packet arrival process is stationary, $P(Y_a = L)$ will equal the fraction of the time axis occupied by CRI's of length L, i.e.,

(4.25) $$P(Y_a = L) = \lim_{n\to\infty} \frac{Ln'_L}{\sum_{i=1}^{\infty} i\,n'_i}, \qquad \text{(a.s.)}.$$

Dividing by n in the numerator and denominator on the right of (4.25) and making use of (4.24) gives

$$P(Y_a = L) = \frac{LP(Y_\infty = L)}{E(Y_\infty)} . \tag{4.26}$$

Multiplying in (4.26) by L and summing over L gives

$$E(Y_a) = \frac{E(Y_\infty^2)}{E(Y_\infty)} , \tag{4.27}$$

where

$$E(Y_\infty^2) \triangleq \lim_{i \to \infty} E(Y_i^2) \quad \text{(a.s.)} \tag{4.28}$$

as follows again from ergodicity.

Now let the random variable Y_d denote the length of the CRI in which the same randomly chosen packet *departs* from the system in the sense of being successfully transmitted, and let X_d be the total number of packets in this CRI. From (4.4) and the fact that in the CORAA a packet departs in the CRI immediately following that in which it arrives, we have

$$E(X_d \mid Y_a = L) = \lambda L. \tag{4.29}$$

Multiplying by $P(Y_a = L)$ and summing gives

$$E(X_d) = \lambda E(Y_a) . \tag{4.30}$$

Next, we note that

$$E(Y_d \mid X_d = N) = L_N \leqslant 2.8867N + \delta_{0N} - 1.8867\delta \tag{4.31}$$

is a simple but rather tight upper bound which follows from (3.21) and a check of the cases $N = 2$ and $N = 3$. Multiplying by $P(X_d = N)$ in (4.31) and summing over N gives

$$E(Y_d) \leqslant 2.8867E(X_d) + P(X_d = 0) - 1.8867P(X_d = 1) . \tag{4.32}$$

But $P(X_d = 1 \mid Y_a = L) = \lambda L e^{-\lambda L} = \lambda L P(X_d = 0 \mid Y_a = L) \geqslant \lambda P(X_d = 0 \mid Y_a = L)$ for all L which implies

(4.33) $$P(X_d = 1) \geq \lambda P(X_d = 0)\,.$$

Using (4.33) in (4.32) gives

(4.34) $$E(Y_d) \leq 2.8867E(X_d) + (1-1.8867\lambda)\ P(X_d = 0)\,.$$

But $P(X_d = 0 | Y_a = L) = e^{-\lambda L} \leq e^{-\lambda}$ for all L so that

(4.35) $$P(X_d = 0) \leq e^{-\lambda}\,.$$

Substituting (4.30) and (4.35) in (4.34) now gives

(4.36) $$E(Y_d) \leq 2.8867\lambda E(Y_a) + (1-1.8867\lambda)e^{-\lambda}$$

provided $\lambda < .53$, which includes all λ in the range of interest as will soon be seen. Inequality (4.36) is our desired tight upper bound on $E(Y_d)$ in terms of $E(Y_a)$.

Similarly, starting from

(4.37) $$L_N \geq 2.8810N - 1 + 2\delta_{0N} - 0.8810\delta_{1N}\,,$$

which follows from (3.21), we note that the same argument that led to (4.32) now gives

$$E(Y_d) \geq 2.8810E(X_d) - 1 + 2P(X_d = 0) - 0.8810P(X_d = 1)\,.$$

Overbounding $P(X_d = 1)$ by $1 - P(X_d = 0)$ gives

(4.38) $$E(Y_d) \geq 2.881E(X_d) - 1.881 + 2.881P(X_d = 0)\,.$$

But $P(X_d = 0 | Y_a = L) = e^{-\lambda L}$; hence multiplying by $P(Y_a = L)$, summing over L, and using Jensen's inequality [8] gives

(4.39) $$P(X_d = 0) \geq e^{-\lambda E(Y_a)}\,.$$

Substituting (4.39) in (4.38) and making use of (4.30) yields

(4.40) $$E(Y_d) \geq 2.881\lambda E(Y_a) - 1.881 + 2.881e^{-\lambda E(Y_a)}\,,$$

which is our desired lower bound on $E(Y_d)$ in terms of $E(Y_a)$.

We now introduce the crucial random variable in a random-access system, namely the *delay* D experienced by a randomly-chosen packet, i.e., the time difference between its arrival at the transmitter and the onset of its successful transmission (so that $D = 0$ when the packet is successfully transmitted beginning at the same moment that it arrives at the transmitter). Now making precise the notion of stability introduced intuitively in Section 4.2, we say that the random-access system is *stable* just when $E(D) < \infty$.

For the CORAA, we first note that

$$\frac{1}{2}E(Y_a) + \frac{1}{2}E(Y_d - 1) \leq E(D) \leq \frac{1}{2}E(Y_a) + E(Y_d - 1) \tag{4.41}$$

as follows from the facts (i) that on the average the randomly-chosen packet arrives at the midpoint of the CRI in progress, and (ii) that at the latest its successful transmission beings in the last slot of its departure CRI, but on the average somewhat beyond the midpoint of the slot starting times as follows from the discussion in Section 2.3. Substituting (4.40) in (4.41) gives the lower bound

$$E(D) \geq \frac{1}{2} + 1.4405\lambda\, E(Y_a) - 1.4405(1 - e^{-\lambda E(Y_a)})\,. \tag{4.42}$$

Similarly, using (4.36) in (4.41) gives the upper bound

$$E(D) \leq \frac{1}{2} + 2.8867\lambda\, E(Y_a) + (1 - 1.8867\lambda)e^{-\lambda} - 1\,. \tag{4.43}$$

From (4.42) and (4.43), it follows that the CORAA is stable if and only if $E(Y_a) < \infty$. We are thus motivated to explore (4.27) more closely.

We first note that $E(Y_\infty^2) \geq E^2(Y_\infty)$ implies by virtue of (4.27) that

$$E(Y_a) \geq E(Y_\infty)\,, \tag{4.44}$$

which in turn, because of (4.17), implies that *the CORAA is unstable for* $\lambda > .3471$.

We now proceed to obtain a rather tight upper bound on $E(Y_a)$. We begin with the bound

$$S_N \leq 8.333N^2 - 2.369N + 1 - 5.964\delta_{1N} + 3.406\delta_{2N} \tag{4.45}$$

which follows from (3.38) and the facts that $S_0 = S_1 = 1$ and $S_2 = 33$. By virtue of the steady-state probabilities in the corresponding Markov chains, we have

$$(4.46)\qquad \begin{aligned} E(Y_\infty^2) &= \sum_{N=0}^{\infty} E(Y^2 | X = N)\; P(X_\infty = N) \\ &= \sum_{N=0}^{\infty} S_N \pi_N . \end{aligned}$$

Thus, multiplying by π_N in (4.43) and summing gives

$$(4.47)\qquad E(Y_\infty^2) \leq 8.333E(X_\infty^2) - 2.369E(X_\infty) + 1 - 5.964\pi_1 + 3.406\pi_2$$

which is a very tight bound as follows from the tightness of (4.45). But from (4.3) and the fact that the mean and variance of a Poisson random variable coincide, we have

$$(4.48)\qquad E(X_{i+1}^2 \,|\, Y_i = L) = \lambda L + (\lambda L)^2 .$$

Multiplying by the steady state probabilities $P(Y_\infty = L)$ and summing over L now gives

$$E(X_\infty^2) = \lambda E(Y_\infty) + \lambda^2 E(Y_\infty^2),$$

which, because of (4.18), can be written as

$$(4.49)\qquad E(X_\infty^2) = E(X_\infty) + \lambda^2 E(Y_\infty^2) .$$

Substituting (4.49) in (4.47) and rearranging gives

$$(1 - 8.333\lambda^2)\, E(Y_\infty^2) \leq 5.964E(X_\infty) + 1 - 5.964\pi_1 + 3.406\pi_2 .$$

Dividing now by $E(Y_\infty) = E(X_\infty)/\lambda$ and using (4.27) yields

$$(4.50)\qquad (1 - 8.333\lambda^2)\, E(Y_a) \leq 5.964\lambda + \frac{1 - 5.964\pi_1 + 3.406\pi_2}{E(Y_\infty)} .$$

We now turn our attention to the second term on the right of (4.50).

First, we observe that

$$E(Y_\infty) = \sum_{i=0}^{\infty} L_N \pi_N$$

$$= 1 + \sum_{i=0}^{\infty} (L_N - 1)\ \pi_i \geq 1 + 4\pi_2 \tag{4.51}$$

because $L_N \geq 1$ for all N and $L_2 = 5$. But (4.51) thus implies

$$\frac{1 + 3.406\pi_2}{E(Y_\infty)} \leq 1, \tag{4.52}$$

which we shall shortly use in (4.50).

To handle the term involving π_1 in (4.50) requires more care. We begin somewhat indirectly by noting that multiplying by $\pi_N = P(X_\infty = N)$ in the upper bound (4.31) on $L_N = E(Y_\infty | X_\infty = N)$ and then summing over N gives

$$E(Y_\infty) \leq 2.8867E(X_\infty) + \pi_0 - 1.8867\pi_1 . \tag{4.53}$$

Overbounding π_0 by $1 - \pi_1$ and using (4.18) gives

$$E(Y_\infty) \leq \frac{1 - 2.8867\pi_1}{1 - 2.8867\lambda} \tag{4.54}$$

provided $\lambda < .3464$. The rather tight bound (4.54) is of some interest in itself. We first use (4.54) only to note that the right side is less than 1 if $\pi_1 > \lambda$; but $E(Y_\infty) \geq 1$ so that by contradiction we conclude that

$$\pi_1 \leq \lambda \tag{4.55}$$

for $\lambda < .3464$. In Section 4.5 we will show that

$$\pi_1 \geq \lambda(1 - \lambda), \qquad \lambda \leq .22 \tag{4.56}$$

which indicates the tightness of (4.55) and (4.56).

Next, we note that (4.54) implies

$$\frac{\pi_1}{E(Y_\infty)} \geq \frac{\pi_1}{1 - \alpha\pi_1}(1 - \alpha\lambda) \tag{4.57}$$

where for convenience we have written

(4.58) $$\alpha = 2.8867\,.$$

The right side of (4.57) increases with π_1 . Thus, using (4.56) in (4.57) gives

(4.59) $$\frac{\pi_1}{E(Y_\infty)} \geqslant \frac{\lambda(1-\lambda)(1-\alpha\lambda)}{1-\alpha\lambda(1-\lambda)}\,, \qquad \lambda \leqslant .22$$

Now using (4.59) for $\lambda \leqslant .22$ and the trivial bound $\pi_1/E(Y_\infty) \geqslant 0$ for $\lambda \geqslant .22$ together with (4.52) in (4.50), we obtain

(4.60) $$E(Y_a) \leqslant \begin{cases} \dfrac{5.964\lambda + 1 - 5.964\lambda(1-\lambda)(1-2.8867\lambda)/[1-2.8867\lambda(1-\lambda)]}{1-8.333\lambda^2}\,, & \lambda \leqslant .22 \\[2ex] \dfrac{5.964\lambda + 1}{1-8.333\lambda^2}\,, & .22 < \lambda < .3464 \end{cases}$$

which is our desired rather tight upper bound on $E(Y_a)$. Because $8.333\lambda^2 = (2.8867\lambda)^2$, (4.60) verifies that *the CORAA is stable for* $\lambda <$ *.3464*, as was anticipated in (4.16).

An entirely similar argument beginning from

(4.61) $$S_N \geqslant 8.300N^2 - 2.403N + 1 - 5.897\delta_{1N} + 3.606\delta_{2N}$$

[which follows from (3.38)], rather than from (4.45), leads to

(4.62) $$(1-8.300\lambda^2)E(Y_a) \geqslant 5.897\lambda + \frac{1-5.897\pi_1 + 3.606\pi_2}{E(Y_\infty)}$$

rather than to (4.50). The tightness of the bounds (4.50) and (4.62) is evident. Underbounding π_2 by 0 and using (4.55) in (4.62) gives

(4.63) $$(1-8.300\lambda^2)E(Y_a) \geqslant 5.897\lambda + \frac{1-5.897\lambda}{E(Y_\infty)}\,.$$

When $\lambda > .1696$ so that $1-5.897\lambda < 0$, we can use the trivial bound $E(Y_\infty) \geqslant 1$ to see that the right side of (4.63) is underbounded by 1. When $\lambda \leqslant 1.696$ we can use (4.54) and (4.56) to show that the right side of (4.63) is underbounded by

$$5.897\lambda + (1-5.897\lambda)(1-2.8867\lambda)/[1-2.8867\lambda(1-\lambda)] .$$

Combining these two bounds, we have

$$E(Y_a) \geqslant \begin{cases} \dfrac{5.897\lambda + (1-5.897\lambda)(1-2.8867\lambda)/[1-2.8867\lambda(1-\lambda)]}{1-8.300\lambda^2}, & \lambda \leqslant .1696 \\ \dfrac{1}{1-8.300\lambda^2}, & .1696 < \lambda < .3464 \end{cases} \quad (4.64)$$

Inequality (4.64) is our desired lower bound on $E(Y_a)$.

Table 4.1 gives a short tabulation of the upper and lower bounds (4.60) and (4.64), respectively, on $E(Y_a)$. The relative tightness of these bounds is perhaps more visible in Figure 4.1. A close review of the bounding arguments suggests that the upper bound (4.60) is a better approximation to $E(Y_a)$ than is the lower bound (4.64).

λ	$E(Y_a)$ upper bound (4.60)	$E(Y_a)$ lower bound (4.64)	E(D) upper bound	E(D) approx. lower bound
0	1	1	1/2	1/2
.05	1.039	1.015	.531	.521
.10	1.179	1.073	.664	.599
.15	1.492	1.216	1.009	.780
.1696	1.696	1.315	1.252	.902
.20	2.165	1.497	1.842	1.200
.25	3.783	2.078	4.03	2.37
.30	11.16	3.952	14.6	9.01
1/3	40.32	12.86	58.2	38.1
.34	82.49	24.68	121.5	80.2
.345	374.5	82.7	559.5	371.9

Table 4.1 : Upper and lower bound on the expected length, $E(Y_a)$, of the CRI in which a randomly-chosen packet arrives versus the throughput λ, and on the expected delay for a randomly-chosen packet, for the CORAA.

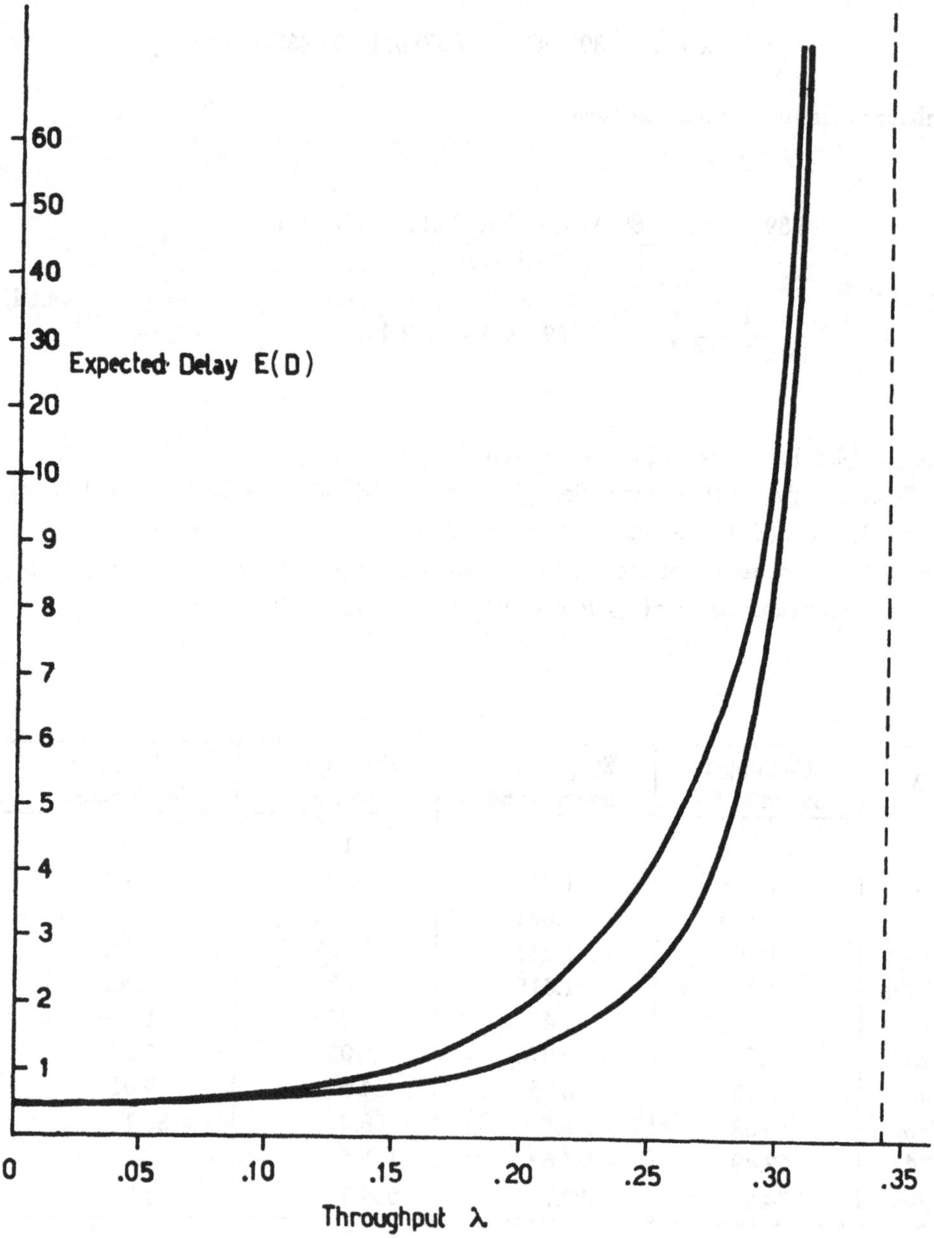

Fig. 4.1: The Upper Bound (4.43) on Expected Delay of a Randomly-Chosen Packet for the CCRA, and an Approximate Lower Bound. (Note change of scale at $\lambda = 10$.)

Using (4.60) in (4.43), we obtain the upper bound on the expected delay, E(D), of a randomly-chosen packet as tabulated in Table 4.1. We could use (4.64) together with (4.42) to get a lower bound on E(D). However, we recall that the tightness of the upper bound (4.60) suggests instead using it together with (4.42) to obtain the "approximate lower bound" tabulated in Table 4.1.
Our purpose is to illustrate that the bounds (4.42) and (4.43) do not significantly differ since their common term $1/2E(Y_a)$ is the dominant one. In Figure 4.1, we have plotted this approximate lower bound on E(D) together with the strict upper bound. As λ is the *throughput* of the system the plot of E(D) versus λ is the *delay-throughput characteristics* of the MCCRAA. Figure 4.1 gives a strict upper bound on this characteristic together with an approximate lower bound which indicates that this upper bound is quite tight.

4.5. Steady-State Probabilities for the CORAA

We have already in (4.21) introduced the steady-state probabilities

$$\pi_N = P(X_\infty = N), \quad N = 0,1,2, \ldots$$

for the CORAA. The equilibrium equations satisfied by these steady-state probabilities are

$$\pi_N = \sum_{n=0}^{\infty} P_{Nn} \pi_n, \quad N = 0,1,2, \ldots \tag{4.65}$$

where P_{Nn} is the transition probability

$$\begin{aligned} P_{Nn} &= P(X_{i+1} = N \mid X_i = n) \\ &= \sum_{L=1}^{\infty} P(X_{i+1} = N \mid Y_i = L)\, P(Y_i = L \mid X_i = n) \\ &= \sum_{L=1}^{\infty} \frac{(\lambda L)^N}{N!} e^{-\lambda L} P(Y_i = L \mid X_i = n) \end{aligned} \tag{4.66}$$

where we have made use of (4.3). In general, these transition probabilities are difficult to calculate. However, because of (3.39), (4.66) for n = 0 and 1 becomes

$$P_{N0} = P_{N1} = \frac{\lambda^N}{N!} e^{-\lambda}, \quad N = 0,1,2, \ldots . \tag{4.67}$$

In Section 3.5, we calculated $P(Y_1 = L \mid X_1 = 2)$. We now use this distribution, given by (3.42), in (4.66) to obtain by summing the resultant series

$$P_{02} = \sum_{m=1}^{\infty} e^{-\lambda(2m+1)}2^{-m}$$

(4.68)

$$= e^{-3\lambda}/(2-e^{-2\lambda}) = A(\lambda) \; ,$$

$$P_{12} = \sum_{m=1}^{\infty} \lambda(2m+1)e^{-\lambda(2m+1)} \, 2^{-m}$$

(4.69)

$$= A(\lambda)\lambda(6-e^{-2\lambda})/(2-e^{-2\lambda})$$

$$\triangleq B(\lambda) \; ,$$

and

$$P_{22} = \sum_{m=1}^{\infty} \frac{1}{2} \lambda^2(2m+1)^2 \, e^{-\lambda(2m+1)} \, 2^{-m}$$

(4.70)

$$= \frac{1}{2} \lambda^2 A(\lambda)[1 + 32/(2-e^{-2\lambda})^2]$$

$$\triangleq C(\lambda) \, .$$

As we shall soon see, these few explicit transition probabilities are quite enough to establish the bound (4.56), which is the main objective of this section.

Making use of (4.67)–(4.7)), we can write (4.65) for N = 1,2 and 3 as

(4.71a) $$\pi_0 = e^{-\lambda}(\pi_0 + \pi_1) + A(\lambda)\pi_2 + \ldots$$

(4.71b) $$\pi_1 = \lambda e^{-\lambda}(\pi_0 + \pi_1) + B(\lambda)\pi_2 + \ldots$$

(4.71c) $$\pi_2 = \frac{1}{2}\lambda^2 e^{-\lambda}(\pi_0 + \pi_1) + C(\lambda)\pi_2 + \ldots$$

where the terms not shown explicitly are of course nonnegative. To proceed further, we need to make use of certain bounds on $E(Y_\infty)$. First, we note that

$$E(Y_\infty) = \sum_{N=0}^{\infty} L_N \pi_N$$

$$\geqslant \pi_0 + \pi_1 + 5\pi_2 + \frac{23}{3}(1 - \pi_0 - \pi_1 - \pi_2)$$

where we have made use of Table 3.1 and the fact that L_N increases with N. Thus, we have

$$3E(Y_\infty) \geq 23 - 20(\pi_0 + \pi_1) - 8\pi_2 \; . \tag{4.72}$$

Discarding the nonnegative higher order terms on the right in (4.71) gives

$$\pi_0 \geq e^{-\lambda}(\pi_0 + \pi_1) + A(\lambda)\pi_2 \tag{4.73a}$$

$$\pi_1 \geq \lambda e^{-\lambda}(\pi_0 + \pi_1) + B(\lambda)\pi_2 \tag{4.73b}$$

$$\pi_2 \geq \frac{1}{2}\lambda^2 e^{-\lambda}(\pi_0 + \pi_1) + C(\lambda)\pi_2 \; . \tag{4.73c}$$

Summing (4.73a) and (4.73b), then rearranging, gives

$$\pi_2 \leq F(\lambda)(\pi_0 + \pi_1) \tag{4.74}$$

where

$$F(\lambda) \triangleq \frac{1-(\lambda+1)e^{-\lambda}}{A(\lambda) + B(\lambda)} . \tag{4.75}$$

Using (4.74) in (4.72) gives

$$3E(Y_\infty) \geq 23 - [20 + 8F(\lambda)]\,(\pi_0 + \pi_1) \; . \tag{4.76}$$

But (4.73c) is equivalent to

$$\pi_2 \geq G(\lambda)(\pi_0 + \pi_1) \tag{4.77}$$

where

$$G(\lambda) \triangleq \frac{1}{2}\lambda^2 e^{-\lambda}/[1 - C(\lambda)] \; . \tag{4.78}$$

Using (4.77) in (4.73b) gives

$$\pi_1 \geq [\lambda e^{-\lambda} + B(\lambda)G(\lambda)]\,(\pi_0 + \pi_1) \; . \tag{4.79}$$

Now using (4.79) in (4.76) gives

$$3E(Y_\infty) \geqslant 23 - \pi_1[20 + 8F(\lambda)]/[\lambda e^{-\lambda} + B(\lambda)G(\lambda)]. \tag{4.80}$$

Inserting the upper bound of (4.54) on $E(Y_\infty)$ into (4.80) and rearraging, we obtain

$$\pi_1 \geqslant (20 - 23a\lambda)/H(\lambda), \qquad \text{if} \quad H(\lambda) > 0 \tag{4.81}$$

where

$$H(\lambda) = \frac{(1-a\lambda)[20 + 8F(\lambda)]}{\lambda e^{-\lambda} + B(\lambda)G(\lambda)} - 3a \tag{4.82}$$

and

$$a = 2.8867.$$

Inequality (4.81) is our desired lower bound on π_1 and is tabulated in Table 4.2. We see that

$$\pi_1 \geqslant \lambda(1-\lambda) \qquad \text{for} \quad \lambda \leqslant .2209 \tag{4.83}$$

which is our desired justification of (4.56).

Proceeding in a similar way, we can derive lower bounds on π_0 and π_2. Using (4.79) in the bound (4.54), we obtain

$$3E(Y_\infty) \leqslant 3\{1 - a[\lambda e^{-\lambda} + B(\lambda)G(\lambda)](\pi_0 + \pi_1)\}/(1 - a\lambda). \tag{4.84}$$

Combining (4.76) and (4.84), we obtain after some rearrangement

$$\pi_0 + \pi_1 \geqslant J(\lambda) \tag{4.85}$$

where

$$J(\lambda) \triangleq \frac{20 - 23a\lambda}{(1-a)[20 + 8F(\lambda)] - 3a[\lambda e^{-\lambda} + B(\lambda)G(\lambda)]} \tag{4.86}$$

λ	$\lambda(1-\lambda)$	Lower Bounds (4.87) π_0	(4.81) π_1	(4.88) π_2	Upper Bounds (4.106) π_0	(4.95) π_1	(4.104) π_2
0	0	1	0	0	1	0	0
.05	.0475	.9506	.04770	.00122	.9510	.04775	.00125
.10	.0900	.9011	.0911	.00485	.9034	.0916	.00516
.15	.1275	.8488	.1299	.0108	.8564	.1319	.0122
.20	.1600	.7872	.1620	.0185	.8094	.1686	.0230
.2209	.1721	.7545	.1721	.0219	.7896	.1828	.0288
.25	.1875	.6909	.1791	.0262	.7620	.2015	.0382
.28	.2016	.5462	.1592	.0263	.7332	.2193	.0497
.30	.2100	.0870	.0272	.0048	.7139	.2303	.0584
.3464	—	—	—	—	.6689	.2533	.0820

Table 4.2 : Upper and Lower Bounds on the Steady-State Probabilities $\pi_N = P(X_\infty = N)$ for the CORAA versus the throughput λ .

provided the denominator of (4.86) is positive. Inserting (4.85) and (4.77) on the right in (4.73a) now gives

$$\pi_0 \geqslant [e^{-\lambda} + A(\lambda)G(\lambda)]J(\lambda) , \tag{4.87}$$

which is our desired lower bound on π_0 . Similarly, using (4.86) in (4.77) gives the bound

$$\pi_2 \geqslant G(\lambda)J(\lambda) . \tag{4.88}$$

A short tabulation of the bounds (4.87) and (4.88) is included in Table 4.2.

Next, we turn our attention to overbounding the steady-state probabilities. Beginning with π_1 , we first note that because

$$xe^{-x} \leqslant e^{-1} , \quad \text{all } x , \tag{4.89}$$

it follows from (4.3) that

$$P(X_{i+1} = 1 | Y_i = L) \leqslant e^{-1} , \quad \text{all } L , \tag{4.90}$$

and hence from (4.66) that

(4.91) $$P_{1n} \leq e^{-1}, \quad \text{all } n.$$

Using (4.91) to overbound the terms for $n \geq 3$ on the right in (4.65) gives

(4.92) $$\pi_1 \leq \lambda e^{-\lambda}(\pi_0 + \pi_1) + B(\lambda)\pi_2 + e^{-1}(1 - \pi_0 - \pi_1 - \pi_2).$$

A simple check shows that $B(\lambda) \leq e^{-1}$ for $0 \leq \lambda \leq 1$ so that (4.77) can be used on the right in (4.92) to give

(4.93) $$\pi_1 \leq e^{-1} - [e^{-1} - \lambda e^{-\lambda} + e^{-1}G(\lambda) - B(\lambda)G(\lambda)](\pi_0 + \pi_1).$$

To proceed further, we need to overbound $\pi_0 + \pi_1$ in terms of π_1. Adding π_1 to both sides of (4.73a) and using (4.77), we obtain

(4.94) $$\pi_0 + \pi_1 \geq \pi_1/[1 - e^{-\lambda} - A(\lambda)G(\lambda)]$$

where we have made use of the fact that $A(\lambda) < 1 - e^{-\lambda}$ for $0 < \lambda \leq 1$. Now using (4.94) in (4.93), we obtain our desired upper bound

(4.95) $$\pi_1 \leq J(\lambda)$$

where

(4.96) $$J(\lambda) = e^{-1}/\{1 + [e^{-1} - \lambda e^{-\lambda} + e^{-1}G(\lambda) - B(\lambda)G(\lambda)]/[1 - e^{-\lambda} - A(\lambda)G(\lambda)]\},$$

provided the expression in wavy brackets is positive. The bound (4.95) is tabulated in Table 4.2.

To obtain upper bounds on π_0 and π_1, we begin by first noting from (4.3) that

(4.97) $$P(X_i = 0 | Y_i = L) = e^{-\lambda L}$$

But, as we saw in Section 3, $X_i \geq 3$ implies $Y_i \geq 5$ because the CRI must contain at least 2 collisions. Thus, (4.94) and (4.97) imply

(4.98) $$P_{0N} \leq e^{-5\lambda}, \quad N \geq 3.$$

Using (4.98) on the right in (4.65) gives

$$\pi_0 \leqslant e^{-\lambda}(\pi_0 + \pi_1) + A(\lambda)\pi_2 + e^{-5\lambda}(1-\pi_0-\pi_1-\pi_2)\,. \tag{4.99}$$

Summing (4.92) and (4.99) gives

$$\pi_0 + \pi_1 \leqslant e^{-1} + e^{-5\lambda} + [(\lambda+1)e^{-\lambda} - e^{-1} - e^{-5\lambda}](\pi_0 + \pi_1) - [e^{-1} + e^{-5\lambda} - A(\lambda) - B(\lambda)]\pi_2\,. \tag{4.100}$$

A simple check shows that $A(\lambda) + B(\lambda) \leqslant e^{-1} + e^{-5\lambda}$ holds for $0 \leqslant \lambda \leqslant .1779$ and $.3179 \leqslant \lambda \leqslant 1$; thus, we can in this range use (4.77) on the right in (4.10) to obtain

$$\pi_0 + \pi_1 \leqslant K_1(\lambda) \quad \text{for} \quad 0 \leqslant \lambda \leqslant .1779 \quad \text{and} \quad .3179 \leqslant \lambda \leqslant 1\,, \tag{4.101}$$

where

$$K_1(\lambda) = (e^{-1} + e^{-5\lambda})/[1 + e^{-1} + e^{-5\lambda} - (\lambda+1)e^{-\lambda} + (e^{-1} + e^{-5\lambda})G(\lambda) - A(\lambda)\,G(\lambda) - B(\lambda)G(\lambda)]\,. \tag{4.102}$$

Similarly, for $.1779 < \lambda < .3179$ we can use (4.74) on the right in (4.100) to obtain

$$\pi_0 + \pi_1 \leqslant K_2(\lambda) \quad \text{for} \quad .1779 < \lambda < .3179 \tag{4.103}$$

where $K_2(\lambda)$ is equal to the right side of (4.102) with $G(\lambda)$ replaced by $F(\lambda)$. It now follows from (4.74) that

$$\pi_2 \leqslant \begin{cases} F(\lambda)K_1(\lambda) & \text{for} \quad 0 \leqslant \lambda \leqslant .1779 \quad \text{and} \quad .3179 \leqslant \lambda \leqslant 1 \\ F(\lambda)K_2(\lambda) & \text{for} \quad .1779 < \lambda < .3179\,. \end{cases} \tag{4.104}$$

This bound is also tabulated in Table 4.2. Finally, we note that using (4.74) in (4.99) gives

$$\pi_0 \leqslant e^{-5\lambda} + (e^{-\lambda} - e^{-5\lambda})(\pi_0 + \pi_1) + [A(\lambda) - e^{-5\lambda}](\pi_0 + \pi_1) \tag{4.105}$$

where we have used the fact that $A(\lambda) \geqslant e^{-5\lambda}$ for $0 \leqslant \lambda \leqslant 1$. Using (4.105) with (4.101) and (4.103) then gives

$$\pi_0 \leqslant e^{-5\lambda} + [e^{-\lambda} - e^{-5\lambda} + F(\lambda)A(\lambda) - F(\lambda)e^{-5\lambda}]K_i(\lambda) \tag{4.106}$$

where

$$\text{(4.107)} \qquad i = \begin{cases} 1 & \text{for} \quad 0 \leqslant \lambda \leqslant .1779 \quad \text{and} \quad .3179 \leqslant \lambda \leqslant 1 \\ 2 & \text{for} \quad .1779 < \lambda < 1 . \end{cases}$$

The bound (4.106) is also tabulated in Table 4.2.

From Table 4.2, we see that our upper and lower bounds on π_0, π_1 and π_2 are so tight for $\lambda \leqslant .15$ as to be virtually equalities, and still reasonably tight for $.15 < \lambda \leqslant .25$. The lower bounds begin to degrade rapidly, however, for $\lambda \geqslant .28$. The chief reason for this is that the upper bound (4.54) on $E(Y_\infty)$ that was used to obtain the lower bounds on π_0, π_1 and π_2 becomes very loose in the region $\lambda \geqslant .28$. To improve our lower bounds, we should make use of (4.50) to obtain an upper bound on $E(Y_\infty)$ that will involve *both* π_1 and π_2, then use this bound in place of (4.54) in the argument that led to the lower bounds on π_0, π_1 and π_2. To obtain even tighter lower bounds, we could begin from an upper bound on $E(Y_\infty)$ in terms of π_1, π_2, and π_3. Etc.

The upper bounds on π_0, π_1 and π_2 above, however, did not utilize any bound on $E(Y_\infty)$ in their derivations. These upper bounds should be virtual equalities for $0 \leqslant \lambda \leqslant .30$ as the inequalities introduced to obtain them are all extremely tight for this range of λ. To tighten these bounds further, we would need to include additional explicit transition probabilities in (4.73).

It should be clear from this section that any finite number of the steady-state probabilities $\pi_0, \pi_1, \pi_2, \ldots$ can be computed to any desired precision by the techniques of this section, if only one has sufficient patience and a good calculator.

4.6. Non-Obvious Random-Access Algorithms with Increased Maximum Stable Throughputs

We saw in Section .4 that the maximum value of the throughputs λ for which the CORAA is stable is (to three decimal places) .347 packets/slot. For the MCORAA, this "maximum stable throughput" is (again to three decimal places) .375 packets/slot. There are many "non-obvious" ways to devise random-access schemes based on the CCRA (or the MCCAA) to increase the maximum stable throughput. Perhaps the most natural way is to "decouple" transmission times from arrival times, as was first suggested by Gallager [9].

Suppose as before that the random-access scheme is activated at time $t = 0$ and that the unit of time is the slot so that the i-th transmission slot is the time interval $(i, i+1]$. But now suppose that a second time increment Δ has been chosen to define *arrival epochs* in the manner that the i-th arrival epoch is the time interval $(i\Delta, i\Delta + \Delta]$. [Note that Δ has units of slots so that $\Delta = 1.5$, for instance, would mean that arrival epochs have length

1.5T seconds, where T is the length of transmission slots in seconds.] Then a very natural way to obtain a random-access algorithm from a collision-resolution algorithm is to use as the first-time transmission rule: *transmit a new packet that arrived during the i-th arrival epoch in the first utilizable slot following the collision-resolution interval (CRI) for new packets that arrived during the (i – 1) arrival epoch.* The modifier "utilizable" reflects the fact that the CRI for new packets that arrived during the (i – 1)-st arrival epoch may end before the i-th arrival epoch. If so, the CRI for the new packets that arrive during the i-th arrival epoch is begun in the first slot that begins after this arrival epoch ends. The "skipped" transmision slots are wasted, and indeed one could improve the random-access algorithm slightly by "shortening" the i-th arrival epoch in this situation – but this complicates both the analysis and the implementation and has no effect on the maximum stable throughput.

The analytical advantage of the above first-time transmission rule is that it completely eliminates statistical dependencies between the resulting CRI's. If X_i denotes the number of new packets that arrive in the i-th arrival epoch and Y_i denotes the length of the CRI for these packets, then $X_0, X_1, X_2, \ldots$ is a sequence of i.i.d. (independent and identically distributed) random variables, and thus so also is $Y_0, Y_1, Y_2, \ldots$. Letting X and Y denote an arbitrary pair X_i and Y_i, we note first that, because the new arrival process is Poisson with a mean of λ packets/slot,

$$P(X = N) = \frac{(\lambda\Delta)^N}{N!} e^{-\lambda\Delta} . \qquad (4.108)$$

Moreover, we have

$$E(Y) = \sum_{N=0}^{\infty} L_N \, P(X = N) \qquad (4.109)$$

and

$$E(Y^2) = \sum_{N=0}^{\infty} S_N \, P(X = N) . \qquad (4.110)$$

In fact, our random-access system is now just a discrete-time queueing system with independent total service times for the arrivals in each arrival epoch. The random-access system is surely *unstable* if

$$E(Y) > \Delta \qquad (4.111)$$

since then the "server" must fall behind the arrivals. Conversely, if

(4.112) $$E(Y) < \Delta$$

and $E(Y^2)$ is finite, then the law of large numbers suffices to guarantee that the average waiting time in the queue will be finite and hence that the random-access system will be stable.

Now consider the use of the CCRA with the above first time transmission rule. From (3.21) and Table (3.1), we have

(4.113) $$L_N \leqslant aN - 1 + 2\delta_{0N} + (2-a)\delta_{0N} + (6-2a)\delta_{2N} + \left(\frac{26}{3} - 3a\right)\delta_{3N}$$

where

(4.114) $$a = 2.8867\,.$$

Substituting (4.113) and (4.108) into (4.109), we find

(4.115) $$E(Y) \leqslant f_a(Z) = aZ - 1 + e^{-Z}[2 + (2-a)Z + (3-a)Z^2 + \left(\frac{13}{9} - \frac{a}{2}\right) Z^3]\,,$$

where we have defined

(4.116) $$Z = \lambda\Delta\,.$$

Using (4.115) in (4.112) [and noting that $E(Y^2)$ is finite as follows from (4.110) and (3.35)], we see that our random-access algorithm will be *stable* for

(4.117) $$\lambda < \sup_{Z>0} \frac{Z}{f_a(Z)} = .4294$$

where the maximizing value of Z is found numerically to be

(4.118) $$Z = \lambda\Delta = 1.147\,.$$

This suggests that the maximum throughput is obtained when the length of the arrival epochs is chosen so that the average number of arrivals is 1.147 – however the maximum in (4.117) is very broad, choosing $\lambda\Delta = 1$ gives a system which is stable for $\lambda < .4277$. Moreover, from (3.21) we see that the inequality (4.113) is reversed if in place of (4.114) we take

(4.119) $$a = 2.8810\,.$$

The condition (4.111) for *instability* of the random-access system is then just

$$\lambda > \sup \frac{Z}{f_a(Z)} = .4295 \tag{4.120}$$

where the maximum is now attained by $Z = \lambda\Delta = 1.148$. Thus, the *maximum stable throughput* of this random-access scheme is based on the CCRA is (to three decimal places) .429 packets/slot, compared to only .347 packets/slot for the CORAA.

If the above first-time transmission rule is used together with the MCCRA, then an entirely similar argument starting from (3.51) shows that this random-access system is *stable* for

$$\lambda < .4622$$

(where the maximizing value is $Z = \lambda\Delta = 1.251$) but is *unstable* for

$$\lambda > .4623 .$$

Thus, the maximum stable throughput is (to three decimal places) .462 packets/slot, compared to only .375 packets/slot for the MCORAA.

A little reflection shows that the increased throughput obtained by using the above first-time transmission rule rather than the "obvious" first-time transmission rule, is that the former avoids the very large initial collisions that occur in the latter when the previous CRI has been so long that many new packets are awaiting first-time transmission.

4.7. Other Variations and the Capacity of the Poisson Random-Access Channel

If one defines the *Poisson Random-Access Channel* by conditions (i), (ii) and (iii) of Section 2.1 together with the specification that for $t > 0$ the new packet process is a stationary Poisson point process with a mean of λ packets/slot, then a quite reasonable definition of its *capacity* is as the supremum, over all realizable random-access algorithms, of the maximum stable throughput obtainable with these algorithms. The maximum stable throughput itself is the supremum of those λ for which the average delay experienced by a randomly-chosen packet is finite when the given random-access algorithm is used. It follows from the results of the previous section that the capacity of the Poisson Random-Access Channel is at least .462 packets/slot.

Note that if a random-access algorithm is stable, i.e., if the average delay for a randomly-chosen packet is finite, then the probability must be unity that a randomly-chosen packet is eventually transmitted successfully. Thus "lossy" random-access algorithms in which there is a nonzero probability that retransmission of a packet is abandoned before it is successfully transmitted are always unstable.

We have specified the branching action within the CCRA and MCCRA to be determined by the results of independent *coin flips* by the various transmitters concerned. It should be clear that we could equivalently have specified this branching to be determined by the *arrival times* of the individual packets at their transmitters. For example, if there has been a collision on the first transmission of the packets in some arrival epoch when the CCRA or MCCRA is used together with the first-time transmission rule of the previous section, then "flipping 0" or "flipping 1" by the colliding transmitters is equivalent to "arriving in the first half of the arrival epoch" or "arriving in the second half of the arrival epoch". If all "coin flips" are so implemented by halving the time interval in question, then the resulting random-access algorithm becomes a first-come-first-served (FCFS) algorithm. (The CORAA and MCRAA likewise become FCFS when this manner of implementing coin flips is used.)

Suppose we use this time-interval-halving method to implement coin flips. As Gallager was first to note [9], if a collision is followed immediately by another collision, then one has obtained no information about the number of packets in the second half of the interval corresponding to the former collision. Thus, the second half interval can be merged into the unexamined portion of the arrival time axis rather than explored as determined by continuation of the collision-resolution algorithm. Using this "trick" with the MCCRA and the first-time transmission rule of the preceding section, Gallager obtained a maximum throughput of .4872 packets/slot (compared to only .462 packets/slot without this "trick".) Mosely [10] refined this approach by optimizing at every step the length of the arrival interval given permission to transmit (which is equivalent to allowing bias in the coin tossed) to obtain a maximum stable througput of .48785; Mosely also gave quite persuasive arguments that this was optimum for "first-come-first-tried" algorithms.

On the other side of the fence, Pippenger [11] used information-theoretic arguments to show that all realizable algorithms are unstable for $\lambda > .744$, and Humblet [12] sharpened this result to $\lambda > .704$. Very recently, Molle [13] used a "magic genie" argument to show that all realizable algorithms are unstable for $\lambda > .6731$ packets/slot.

Thus, the capacity of the Poisson Random-Access Channel lies somewhere between .48785 packets/slot and .6731 packets/slot, and no more can be said with certainty at this writing. Beginning with Capetanakis [3], most workers on this problem have conjectured that the capacity is 1/2 packet/slot. In any event, it has recently appeared much easier to reduce the upper bound on capacity then to increase the lower bound. And it is no longer defensible for anyone to claim that 1/e is "capacity" in any sense.

5. EFFECTS OF PROPAGATION DELAYS, CHANNEL ERRORS, ETC.

In Section 2.1, we stated the idealized conditions under which the previous analysis of collision-resolution and random-access algorithms was made. We now show how the ideal-case analysis can be easily modified to include more realistic assumptions and also to take advantage of additional information sometimes available in random-access situations.

5.1. Propagation Delays and the Parallel Tree Algorithm

Assumptions (ii) and (iii) in Section 2.1 stipulated that, immediately at the end of each slot, each transmitter received one of 3 possible messages, say "ACK" or "NAK" or "LAK", indicating that one packet had been successfully transmitted in that slot or that there had been a collision in the slot or that the slot had been empty, respectively. This assumption is appropriate, however, only when the round trip transmission time is much smaller than one slot length. Suppose that the *round-trip propagation delay* time plus the transmission time for the feedback message (ACK or NAK or LAK) is actually D_r slots. (If the propagation delay varies among the transmitters, then D_r is the maximum such delay.) In this case, the result of transmission in slot i can govern future transmissions no earlier than in slot i + d where

$$d = \lceil D_r \rceil + 1 \tag{5.1}$$

and where $\lceil x \rceil$ denotes the smallest integer equal to or greater than x. For instance, $D_r > 2.3$ slots would imply d = 4 slots.

The simplest way conceptually to extend our former $D_r = 0$ analysis to the case $D_r > 0$ is to treat the actual random-access channel as d *interleaved zero-propagation delay channels.* Slots 0, d, 2d, . . . of the actual channel form slots 0,1,2, . . . of the first interleaved channel. Slots 1, d + 1, 2d + 1, . . . of the actual channel form slots 0,1,2, . . . of the second interleaved channel; etc. Whatever *random-access algorithm* is chosen *is independently executed on each of the d interleaved channels.*

When the CORAA [or the MCORAA] is used on the individual interleaved channels, perhaps the most natural *traffic assignment rule* for new packets is that, if the new packet arrives at its transmitter during slot i − 1, it is assigned to the next occurring interleaved channel, i.e., to the interleaved channel corresponding to slots, i, i + d, i + 2d, ... We now suppose that this assignment rule is adopted. If the expected delay for a randomly-chosen packet for the case $D_r = 0$ and Poisson traffic with a mean of λ packets/slot is E(D), then the expected delay $E(D_i)$ for a randomly-chosen packet for the interleaved scheme and Poisson traffic with the same mean is just

$$E(D_i) = dE\left(D - \frac{1}{2}\right) + \frac{1}{2} \tag{5.2}$$

This follows from the facts that each interleaved channel still sees Poisson traffic with a mean of λ packets/slot, and that on the average a new packet waits 1/2 slot before becoming active in the algorithm for the interleaved channel and thus waits E(D – 1/2) further transmission slots on the interleaved channel before the start of its successful transmission. Using (5.2), we can easily convert the delay-throughput characteristic for the $D_r = 0$ case to that for the interleaved scheme.

Despite the naturalness of the above new packet assignment rule for the interleaved CORAA (or the MCORAA), it is obviously inferior to the rule: assign a newly arrived packet randomly to one of the interleaved channels having no collision-resolution interval (CRI) in progress, if any, and otherwise to the next occurring CRI. But it appears difficult to calculate the resulting improvement in performance.

When arrival epochs are distinguished from transmission slots as suggested in Section 4.6, then a good traffic assignment rule for the interleaved channels is: transmit a new packet that arrived during the i-th arrival epoch in the next slot that occurs after new packets in the (i–1)-st arrival epoch have been initially transmitted and is from an interleaved channel with no CRI in progress.

In the infinitely-many-sources-generating-Poisson-traffic model, there is never a queue at the individual transmitters. In the practical case, however, a transmitter may receive one or more additional new packets before successfully transmitting a given new packet. In the interleaved case, this circumstance can be exploited to reduce the expected delay by assigning the additional new packets to other interleaved channels so that one transmitter can be actively processing several packets at once. Of course, this means also that these packets might be successfully transmitted in different order from their initial arrivals.

The interleaved scheme for coping with propagation delay is probably the simplest to implement as well as to analyze. It does, however, achieve this simplicity at the price of some increased delay. An alternative to interleaving suggested by Capetanakis |3,4| is to re order the transmissions in the CCRA so as to process all the nodes at each level in the tree before going on the next level. For instance, for the situation illustrated in Figures 2.1 and 2.2, the order of the slots would effectively be permuted to 1,2,5,3,4,6,7,8,11,9,10. Note for instance that the transmitters colliding in slot "5" are idle for the following two slots so that if d = 1 or d = 2 there would be no wait required before they could proceed with the algorithm. Capetanakis called this scheme the *parallel tree algorithm* to contrast it with the "serial tree algorithm" (or CCRA in our terminology).

The parallel tree algorithm appears attractive for use in random-access systems where the propagation delays are large. Thus, it seems worthwhile to show here that it can be implemented just as easily as in the CCRA. Just as was done in Section 2.3 when considering the CCRA, we suppose that each transmitter keeps two counters (which we now call C_1 and C_2), the first of which indicates by a zero reading that he should transmit and the second of which indicates the number of additional slots that have already been allocated for further transmissions within the CRI in progress. The parallel tree algorithm differs from the CCRA only in the respect that, after a collision, the colliding transmitters go to the end of the waiting line rather than remaining at the front. Thus, with the stipulation that all transmitters set C_2 to 1 just before the CRI begins, we can implement the parallel tree algorithm by the following modification of the implementation given for the CCRA in Section 2.3.

Parallel Tree Algorithm Implementation: When a transmitter flips 0 or 1 following a collision in which he is involved, he sets counter C_1 equal to $C_2 - 1$ or C_2, respectively, then increments C_2 by 1. After all other slots, he decrements C_1 by 1, and transmits in the next slot when C_1 reaches 0. After noncollision slots, he decrements C_2 by 1; after collisions in which he was not involved, he increments C_2 by 1. When C_2 reaches 0, the CRI is complete.

Notice that, after a collision, those colliding transmitters who flip 0 will transmit again exactly C_2 slots later, i.e., in slot $i + C_2$ when the collision was in slot i. Thus, the following artifice suffices to ensure that the ACK, NAK or LAK reply from the receiver about some slot will reach every transmitter before it is called upon to act on the result of transmissions in that slot.

Parallel Tree Algorithm Implementation When Propagation Delays are Large and Interleaving is Not Used: Same as the ordinary implementation given above except that, when a collision occurs with $C_2 < d$, all transmitters immediately reset C_2 to d before proceeding with the other actions required by the algorithm. We note that, when C_2 is reset to d, $d - C_2$ slots will be "wasted" by the algorithm, i.e., these slots will necessarily contain no transmissions. These slots could of course be used to begin (or to continue with) another collision resolution interval – but this would so complicate the implementation that one would probably be better advised to use the interleaved approach if the efficient use of all channel slots is so important as to warrant such complexity.

5.2. Effects of Channel Errors

Assumption (i) and (ii) of Section 2.1 stipulated that, except for collisions in the forward channel, both the forward and feedback channels in the random-access model were "noiseless". In other words, after each transmission slot, all transmitters are correctly informed as to whether the slot was empty ("LAK"), or contained one packet ("ACK"), or contained a collision ("NAK"). We now consider the more realistic situation where channel noise can affect the transmissions on the forward channel or feedback channel, or both.

The basic model for our error analysis is the discrete memoryless channel shown in Figure 5.1. The input to this channel is the actual status of the given transmission slot, i.e.,

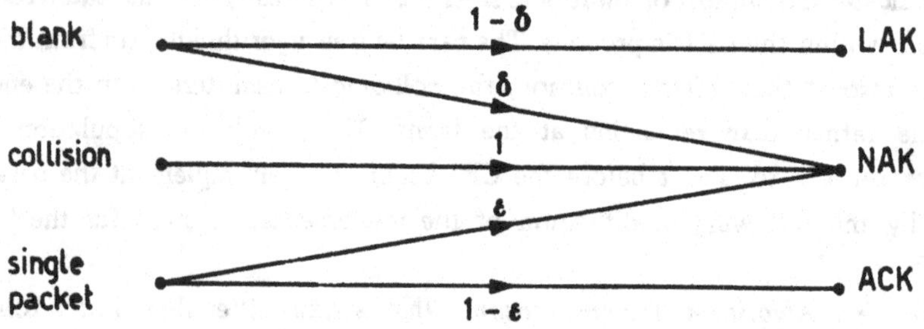

Fig. 5.1: Model for Analysis of Channel Error Effects.

"blank" or containing a "single packet" or containing a "collision". The output is the actual feedback message reaching the transmitters, i.e., "LAK" or "ACK" or "NAK". In practice, each packet transmitted would be encoded with a sufficiently powerful error-detecting code that the probability would be so negligible that the common receiver would incorrectly identify either a "blank" or a "collision" as a "single packet" because of errors in the forward channel. We also presume that the transmitters will interpret any message garbled by noise on the feedback channel as a "NAK", and that these feedback messages would also be coded to make negligible the probability that the transmitters would incorrectly identify either a feedback "LAK" or a "NAK" as an "ACK", or would incorrectly identify either a feedback "ACK" or "NAK" as an "ACK", because of errors on the feedback channel. Thus, it is realistic to assume that the only types of errors that can occur in our random-access system are those that result in the transmitters interpreting the feedback message as "NAK" when in fact the slot had been "blank" or had contained a "collision". These two types of errors are indicated by the two nondirect transitions shown in the error model of Figure 5.1. We write δ to denote the probability that the transmitters will incorrectly conclude that a blank transmission slot contained a collision, and ε to denote the probability that they will incorrectly conclude that a transmission slot with a single packet contained a collision. Note that δ (as well as ε) accounts for error effects on *both* the forward and feedback channels. A blank slot could, because of errors on the forward channel, reach the common receiver as a "garble", thus eliciting a "NAK" message on the feedback channel; or a blank slot could be correctly identified by the common receiver but the subsequent "ACK" message, because of noise on the feedback channel, might reach the transmitters as a "garble" that they are forced to interpret as a "NAK" – both of these cases are included in the transition from "blank" to "NAK" for the error model of Figure 5.1.

In a realistic random-access situation, one would expect δ to be much smaller than ε, as only one bit error in a packet on the forward channel could cause a "single packet" slot to result in a "NAK" by the common receiver (unless some error-correction were employed in addition to error detection). Moroever, for the system to be useful, one requires $\varepsilon << 1$, for otherwise the throughput would be low because of the need for transmitting a packet many times before it is successfully received even if a "genie" were to assist the transmitters to schedule their transmissions so that collisions never occurred. Thus, typically, one would anticipate the inequality

$$\delta << \varepsilon << 1 . \tag{5.3}$$

However, we need not impose this requirement in order to analyze the effect of errors on the Capetanakis collision-resolution algorithm (CCRA).

We begin our analysis somewhat indirectly by first extending our analysis of the CCRA for the *error-free case.* Recalling the definitions of Section 3.1, we now further define Y_b to be the number of blank slots in the CRI, Y_s to be the number of slots with a single packet, and Y_c to be the number of slots with collisions. Referring to the tree diagram (such as in Figure 2.2) for the CRI, we see that $Y_b + Y_s$ is the number of terminal nodes whereas Y_c is the number of intermediate nodes. But, as we noted in Section 2.3, a binary rooted tree always has exactly one more terminal node than intermediate nodes so that

$$Y_b + Y_s = Y_c + 1 \tag{5.4}$$

We next define

$$B_N = E(Y_b | X = N) \tag{5.5}$$

and

$$C_N = E(Y_c | X = N) . \tag{5.6}$$

Now we observe that

$$E(Y_s | X = N) = N \tag{5.7}$$

since each of the N packets in the first slot of the CRI is successfully transmitted exactly once in the CRI. But, of course

$$Y = Y_b + Y_s + Y_c \tag{5.8}$$

so that, upon taking the expectation conditioned on X = N, we have

(5.9) $$L_N = B_N + N + C_N \, .$$

Similarly, (5.4) yields

(5.10) $$B_N + N = C_N + 1 \, .$$

Solving (5.9) and (5.10), we obtain

(5.11) $$B_N = (L_N + 1 - 2N)/2$$

which is a fundamental relationship for the CCRA in the error-free case.

We are now ready to grapple with errors. In Figure. 5.2, we show the effect in the

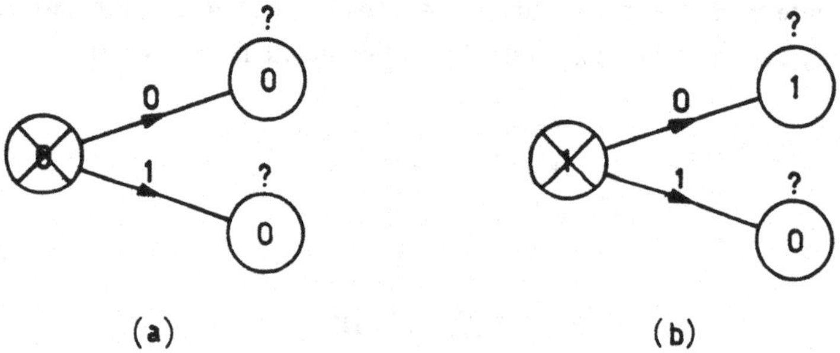

Fig. 5.2: Immediate effect of (a) a blank-to-NAK error and (b) a single-packet-to-NAK error in the CCRA.

resulting tree diagram for the CRI of a blank-to-NAK and a single-packet-to-NAK error on the operation of the CCRA. Here, a cross inside a node indicates that, because of channel errors, that node will be interpreted as a collision slot by the transmitters. The question mark above the subsequent nodes indicates that we do not yet know whether, because of further channel errors, that node will also be interpreted as a collision slot rather than correctly as a blank slot or single packet slot. Thus, a blank-to-NAK error has the immediate effect of adding two additional blank slots to the CRI, whereas a single-packet-to-NAK error adds one blank slot and one single packet slot.

We now define the random variable t_b to be the number of slots required for the successful transmission of a given blank slot when errors are present, i.e., the number of slots until all spurious "collisions" have been resolved and found to have been actually blank slots. From Figure 5.2(a), we see that

$$E(t_b) = 1(1-\delta) + [1 + 2E(t_b)]\delta , \tag{5.12}$$

because $t_b = 1$ when the blank slot initially results in a correct LAK reply as happens with probability $1 - \delta$, but on the average $2E(t_b)$ further slots will be required if the blank slot initially results in an erroneous NAK reply as happens with probability δ. From (5.12) we find

$$E(t_b) - 1 = \frac{2\delta}{1-2\delta} , \quad \delta < \frac{1}{2} \tag{5.13}$$

where we have focused interest on $E(t_b) - 1$ which is the *expected number of extra slots added because of channel errors to the CRI for each blank slot in the error-free case.*

Similarly, we define t_s to be the number of slots required for the successful transmission of a given single packet slot when errors are present. Referring to Figure 5.2(b), we see that

$$E(t_s) = 1(1-\varepsilon) + [1 + E(t_b) + E(t_s)]\varepsilon . \tag{5.14}$$

Substituting (5.13) into (5.14) and solving, we find

$$E(t_s) - 1 = \frac{2\varepsilon(1-\delta)}{(1-2\delta)(1-\varepsilon)} , \quad \delta < \frac{1}{2} , \ \varepsilon < 1 \tag{5.15}$$

which is the *expected number of extra slots added because of channel errors to the CRI for each single-packet slot in the error-free case.*

It now follows immediately from (5.13) and (5.15) that

$$E(Y|X = N, \text{errors}) = L_N + B_N \frac{2\delta}{1-2\delta} + N \frac{2\varepsilon(1-\delta)}{(1-2\delta)(1-\varepsilon)} . \tag{5.16}$$

Using (5.11) and (5.16) and simplifying, we find

$$E(Y|X = N, \text{errors}) = \frac{1-\delta}{1-2\delta} L_N + \frac{2(\varepsilon-\delta)}{(1-2\delta)(1-\varepsilon)} N + \frac{\delta}{1-2\delta} . \tag{5.17}$$

The fundamental relation (5.17) now permits us to make use of the tight bounds on L_N developed in Section 3.3 to obtain tight bounds on the expected CRI length in the presence of errors.

In particular, it follows from (3.21) and (5.17) that

$$(5.18) \qquad E(Y|X = N, \text{errors}) \leq \left[\frac{2.8867(1-\delta)}{1-2\delta} + \frac{2(\varepsilon-\delta)}{(1-2\delta)(1-\varepsilon)}\right] N - 1$$

for all $N \geq 4$. The stability anaysis of Sections 4.3 and 4.4 can now immediately be invoked to assert that *the CORAA is stable in the presence of channel errors provided that*

$$(5.19) \qquad \lambda < \left[\frac{2.8867(1-\delta)}{1-2\delta} + \frac{2(\varepsilon-\delta)}{(1-2\delta)(1-\varepsilon)}\right]^{-1}$$

where λ is the throughput, i.e., the average number of new packets per slot, of the Poisson traffic. Moreover, we know from the tightness of the upper bound (3.21) on L_N that the right side of (5.19) is very nearly equal to the maximum stable throughput.

In Table 5.1, we show the tight lower bound on the maximum stable throughput given by the right side of (5.19) over a wide range of ε and δ. The values of ε and δ have not been chosen to correspond to a practical situation [cf. (5.3)], but rather to demonstrate that *the CORAA is remarkably insensitive to channel errors.* Even for the practically extreme values $\varepsilon = \delta = .1$, the maximum stable throughput is still 90% of its value in the error-free case.

In light of the insensitivity of the CORAA to channel errors, it seems quite surprising that the MCORAA is extremely sensitive to such errors, as we now proceed to show. We will say that a collision-resolution algorithm suffers from *deadlock due to channel errors*, if, because of a finite number of errors, the resulting CRI never terminates although its first slot contains only a finite number of packets. Clearly, a collision-resolution algorithm with such deadlock would be a disastrous choice for inclusion within a random-access algorithm.

In Figure 5.3, we give an example to show that *the modified Capetanakis collision-resolution algorithm (MCCRA) suffers from deadlock due to channel errors.* In this example, the first slot of the CRI is actually blank but, because of a blank-to-NAK error, is construed by the transmitters to have contained a collision. They all wait for those colliding transmitters who flip 0 to send in slot 2, which of course must then be blank. But the MCCRA now directs the colliding transmitters to skip what erroneously appears to be a certain collision among the colliding transmitters who flipped one. These latter are directed to flip again with those who now flip 0 then transmitting in slot 3, which of course must again be blank, etc. A single blank-to-NAK error thus results in an infinitely long CRI even though there are actually no packets to be transmitted!

ε	δ	Lower Bound (5.19) on Maximum Stable Throughput
0	0	.3464
.01	0	.3440
.10	0	.3217
.20	0	.2953
.50	0	.2046
.80	0	.0919
0	.01	.3453
0	.10	.3336
0	.20	.3142
0	.40	.2146
0	.45	.1454
.10	.10	.3079
.20	.20	.2598
.30	.30	.1980
.40	.40	.1155
.10	.01	.3205
.10	.02	.3193
.20	.01	.2940
.20	.02	.2928

Table 5.1 : The Lower Bound (5.19) on the Maximum Stable Throughput of the CORAA for Various Values of ε (the single-packet-to-NAK error probability) and δ (the blank-to-NAK error probability).

Naturally one could "doctor" the MCCRA to avoid the above illustrated deadlock, say by specifying that no more than 2 successive blank-skip slots will be permitted. But this further complicates its implementation and also reduces its maximum stable throughput under error-free conditions. When one reflects that the maximum stable throughputs of the CORAA and the MCORAA are .347 and .375, respectively [and that, for the non-obvious first time transmission rule of Section 4.6, are .430 and .462 for the CCRA and MCCRA, respectively] in the error-free case, one can hardly escape the conclusion that the slightly increased throughput of the MCCRA comes at too great a price in increased sensitivity to channel errors compared to the CCRA.

Returning to consideration of the CCRA, we now address some possible objections to our error analysis. One might argue that we should have allowed collision-to-ACK or collision-to-NAK errors in our analysis, even though they may have very small probabilities.

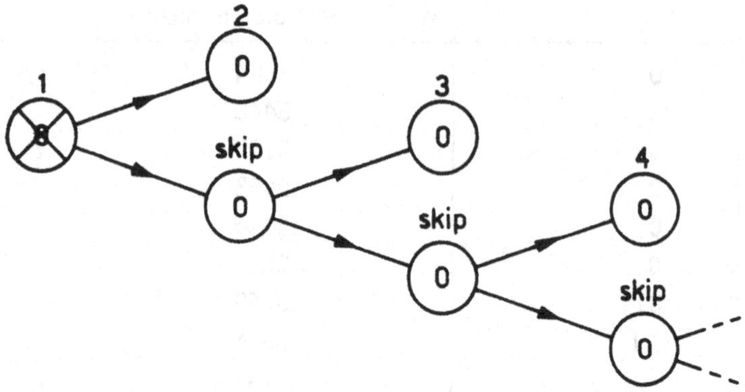

Fig. 5.3: Example of deadlock due to channel errors in the MCCRA.

But a collision-to-ACK error will cause all the colliding transmitters to believe that their packets were successfully transmitted and thus will actually shorten the CRI. Naturally these packets are forever lost, unless some accounting is performed at the destination and a repeat-request sent back – but this is the usual and unavoidable way that any communication system must occasionally fail, and is not related to random-access issues. On the other hand, a collision-to-LAK error will be immedaitely recognized by all the colliding transmitters who then presumably would save their packets for transmission in the next CRI. The only two conceivable types of errors that remain unconsidered are blank-to-ACK errors, which cause no disruption at all in the CCRA, and single-packet-to-LAK errors, which again will each be recognized by the transmitter in question so that the packet can be transmitted in the next CRI. The heartening conclusion is that *the CCRA is extremely insensitive to channel errors of every possible sort.*

Finally, one could object that we have not allowed the possibility that a feedback message from the common receiver could be correctly received by some transmitters but incorrectly understood as a NAK by the rest. It is tedious, but not difficult, to enumerate all types of such errors and to find again that none does significant damage to the operation of the CCRA. The robustness of the CCRA in the face of every sort of error really stems from the fact that *a priori* it assumes always that there *may* be a collision in the next slot. If an unauthorized packet appears in this slot because of channel errors on past transmissions, it merely becomes a part of the collision to be resolved from that point onwards. This packet's shady past does not preclude a bright future. The CCRA even rides through periods where some transmitters lose track of the correct endpoints of the CRI's because eventually from this point onwards there will be more non-collision slots than collision slots so that all transmitters will again agree on the endpoint of a CRI. The MCCRA, however, is prone to disaster from channel errors because it rashly reaches the conclusion that a collision is *certain* to occur in some slot from imperfect information garnered from previous slots.

5.3. Carrier-Sensing

In some random-access situations, particularly in packet-radio networks, it is possible for the transmitters to "hear" that a transmission slot is empty or, when sending a packet, to "hear" that interfering signals are present. In either case, the transmitters can then abruptly terminate the otherwise unproductive slot. Such techniques are generally known as "carrier-sensing" [14], and we now analyze their effectiveness in conjunction with the Capetanakis collision-resolution algorithm (CCRA).

Let θ_b be the fraction of a slot required for all transmitters to detect that a transmission slot is empty (i.e., "blank"), and let θ_c similarly be the fraction of a slot required to detect that a transmission slot contains a collision. Then, if *carrier-sensing* (c–s) is exploited to terminate blank and collision slots as soon as they are detected by all transmitters, the effect is merely to reduce the length of each blank or collision slot by a factor of $1-\theta_b$ or $1-\theta_c$, respectively. Thus, it follows immediately that the expected length of a collision-resolution interval (CRI) for the CCRA is given by

$$E(Y|X=N,c-s) = L_N-(1-\theta_b)B_N-(1-\theta_c)C_N \tag{5.20}$$

where B_N and C_N are defined in the previous section.

Solving (5.9) and (5.10), we find

$$C_N = (L_N-1)/2\,. \tag{5.21}$$

Now using (5.11) and (5.21) in (5.20), we obtain

$$E(Y|X=N,c-s) = L_N(\theta_b+\theta_c)/2+N(1-\theta_b)+(\theta_b-\theta_c)/2\,, \tag{5.22}$$

which is a fundamental relation. Just as for the error analysis of the preceding section, we can now exploit in (5.22) the tight bounds on L_N developed in Section 3.3.

Using (3.21) in (5.22), we find

$$E(Y|X=N,c-s) \leqslant [1.4434(\theta_b+\theta_c)+1-\theta_b]N-\theta_c \tag{5.23}$$

for all $N \geq 4$. The stability analysis of Sections 4.3 and 4.4 can now immediately be invoked to assert that *the CORAA is stable provided that*

$$\lambda < [1.4434(\theta_b + \theta_c) + 1 - \theta_b]^{-1}, \tag{5.24}$$

where λ is the throughput, i.e., the average number of new packets per (unshortened) slot, of the Poisson traffic. Again we know from the tightness of the upper bound (3.21) on L_N that the right side of (5.24) is virtually equal to the maximum stable throughput.

In Table 5.2, we show the tight lower bound on the maximum stable throughput of the CORAA given by the right side of (5.24) over a wide range of θ_b and θ_c. Not surprisingly, the maximum stable throughput is unity for $\theta_b = \theta_c = 0$, since then blank slots and collision slots are reduced to zero length so that every actual transmission slot can be used for successful transmissions. More significantly, Table 5.2 shows that early detection

θ_b	θ_c	Lower Bound (5.24) on Maximum Stable Throughput
0	0	1
.1	.1	.841
.2	.2	.726
.5	.5	.515
.7	.7	.431
1	1	.346
1	.9	.365
1	.7	.408
1	.5	.462
1	.2	.577
1	0	.693
.5	1	.375
.2	1	.395
0	1	.409

Table 5.2 The Lower Bound (5.24) on the Maximum Stable Throughput of the CORAA for Various Values of θ_b (fraction of slot required to detect no transmission) and θ_c (fraction of slot required to detect a collision).

of collisions is far more important than early detection of blank slots. For instance, the maximum stable throughput for instant detection of blank slots ($\theta_b = 0$) is only .409 when $\theta_c = 1$, compared to .346 when $\theta_b = \theta_c = 1$; but this same increased stable throughput can be attained by the rather late detection of collisions, $\theta_c = .7$, when $\theta_b = 1$.

5.4. Etc.

Virtually any of the "tricks", such as making short "reservations" on the channel for the later transmission of long "messages" [15], that have been suggested for increasing the throughput of Aloha-like random-access systems can also be incorporated into random-access systems based on the Capetanakis collision-resolution algorithm (CCRA). Our aim in the previous section has been to illustrate for one such "trick", viz., carrier-sensing, how our previous analysis of the CCRA and the CORAA can readily be extended to calculate the resulting enhancement of the random-access system without any appeal to "statistical equilibrium". The reader should have no difficulty in making the appropriate extensions for the other tricks in the bag of the random-access system designer.

6. SUMMARY AND HISTORICAL CREDITS

In the previous text, we have given a rather thorough analysis of the Capetanakis collision-resolution algorithm (CCRA) and its use within random-access systems. We have stressed the algorithmic properties of the CCRA itself so that our main results are independent of traffic statistics. When calculating the performance of random-access systems based on the CCRA, however, we have generally used the usual Poisson traffic model. As we have repeatedly emphasized, all of our calculations have been mathematically rigorous; in particular, they have been devoid of evasive appeals to the assumption of "statistical equilibrium" that has been all too pervasive in the random-access literature. Finally, we have extended our analysis to include the effects of channel errors and propagation delays, and to calculate the benefits from "carrier-sensing".

In many places in the text, we have pointed out the original source of the result being discussed, but we now attempt to fill in as many omissions of such credit as possible.

The pioneering work of Capetanakis [3–5] has of course been the main source and inspiration of this paper. It is doubtful that any scheme so elegantly simple as the CCRA algorithm has no prior roots in the literature. Some such roots of the CCRA can be seen in the polling algorithm proposed by Hayes [16], but the fundamental concept of a collision-resolution algorithm seems to have found no expression prior to the work of Capetanakis.

The author suggested to Capetanakis the modification to eliminate certain collisions that we have called the MCCRA – not knowing at the time that it leads to deadlock with channel errors. Quite interestingly, this MCCRA was independently but somewhat later discovered by Tsybakov and Mihailov [7] who bypassed the more robust algorithm. These latter authors, however, were the first to use a recursive analysis in their study of the MCCRA and MCORAA. The extremely tight and systematic bounds of Section 3 are due to the author's student, M. Amati, and will form part of his doctoral dissertation. The analysis of Sections 4.1 through 4.5 is largely the joint work of Amati and the author, but the maximum stable throughputs calculated there had already been found by Capetanakis [3,4] using Chernoff bounds.

Capetanakis [3,4] originally gave a modification of his algorithm, based on dynamically varying the degree of the root node in the tree, that yields the same maximum stable throughputs of .429 and .462 as calculated in Section 4.6; our approach there, which was based on Gallager's trick of divorcing the arrival time axis from the transmission time axis, is virtually equivalent to, but conceptually simpler than, Capetanakis' "dynamic tree algorithm". The implementation of the parallel tree algorithm suggested in Section 5.1 is due to the author, as is the analysis of Sections 5.2 and 5.3, and as are any errors that the reader may have found.

ACKNOWLEDGEMENT

The support by the U.S. Office of Naval Research (Contract No. N00014-78-C-0778) of the original research by M. Amati and the author reported herein is gratefully acknowledged.

22nd October, 1980

POSTSCRIPT

The appearance of the English translation of [7] has shown that our deficiency in Russian resulted in our giving insufficient credit to this work of Tsybakov and Mikhailov. Not only did they independently describe and analyze the MCCRA and the MCORAA as stated above in Section 6, but they also and independently described the CCRA before dismissing it as "inferior" to the MCCRA. The CCRA, or "serial tree algorithm", should thus rightly be termed the Capetanakis-Tsybakov-Mikhailov collision-resolution algorithm (CTMCRA), and we urge the reader to adopt this terminology. Similarly, the MCCRA should rightly be termed the Capetanakis-Massey-Tsybakov-Mikhailov collision-resolution algorithm (CMTMCRA) [where the immodesty of including our name for a minor modification to the major work of Capetanakis owes to the fact that Capetanakis failed to notice the existence of certain-to-occur collisions in his algorithm.] Capetanakis, however, seems to have been the sole originator of the "parallel tree algorithm" described in Section 5.1.

During a recent visit in Moscow, we learned of several recent and interesting results by Soviet researchers on collision-resolution algorithms. In their oral presentation, "Methods of Random Multiple Access", at the International Symposium on Information Theory, Tbilisi, July 3-7, 1979, B.S. Tsybakov, M.Y. Berkovskii, N.D. Vvedenskaya, V.A. Mikhailov and S.P. Fedorev independently proposed the "trick" of decoupling transmission time from arrival time [cf. Section 4.6] and also performed an optimization over three parameters to obtain a random-access algorithm for the Poisson case with a maximum stable throughput of .4877 packets/slot. The most significant new result, however, in our opinion is that contained in the paper, "Stack Algorithm for Random Multiple Access", by Tsybakov and Vvedenskaya that will appear in the July-October 1980 number of *Problemy Peredachi Informatsii.* Tsybakov and Vvedenskaya consider a "continuous-entry" random-access algorithm in which transmitters always send new packets in the slot immediately following their arrival, then resolve any collisions using the CMTMCRA. (In other words, transmitters with new packets do not wait for the CRI in progress to end before joining the contending set of transmitters). They showed that the maximum stable throughput for the Poisson case is .384 packets/slot. These analytical results are corroborated by simulations that we conducted in 1978 for the same algorithm, but our subsequent analytical efforts were fruitless. The practical significance of the continuous-entry feature is that it eliminates the undesirable necessity for transmitters to monitor the channel during periods when they have no packet to send. It seems surprising to us that, as reference to (3.51) above shows, the continuous-entry algorithm outperforms the "obvious random-access algorithm" incorporating the CMTMCRA. But even more surprising to us is an as-yet-unwritten result due to Vvedenskaya showing that, for the CMTMCRA, L_N/N does not approach a limit as N tends to infinity, but rather oscillates in he fifth or sixth significant decimal digit. Presuming this state of affairs also holds the CTMCRA, we must caution the reader to view (3.31) above with suspicion. While the plausibility argument for (3.31) has some merit, our carelessly suggested method of proof could not succeed. Fortunately, no important use of (3.31) was made above.

REFERENCES

[1] N. Abramson, "The ALOHA system – another alternative for computer communications", *AFIPS Conf. Proc.*, Fall Joint Comp. Conf., vol. 37, pp. 281-285, 1970.

[2] L.G. Roberts, "ALOHA packet system with and without slots and capture", ASS Note 8, ARPA Network Infor. Ctr., Stanford Res. Inst., Stanford, CA, June 1972.

[3] J.I. Capetanakis, "The multiple access broadcast channel: protocol and capacity considerations", Ph.D. Thesis, Course VI, Mass. Inst. Tech., Cambridge, MA, August 1977; orally presented at IEEE Int. Symp. Info. Th., Ithaca, NY, October 10-14, 1977.

[4] J.I. Capetanakis, "Tree algorithms for packet broadcast channels", *IEEE Trans. Info. Th.*, vol. IY–25, pp. 505-515, September 1979.

[5] J.I. Capetanakis, "Generalized TDMA: The multi-accessing tree protocol", *IEEE Trans. Comm.*, vol. COM–27, pp. 1476-1484, October 1979.

[6] R.G. Gallager, Private Communication, January 1978.

[7] B.S. Tsybakov and V.A. Mihailov, "Slotted multiaccess packet-broadcasting feedback channel, "*Problemy Peredachi Informatsii*, vol. 14, pp. 32-59, October-December 1978.

[8] A.J. Viterbi and J.K. Omura, *Principles of Digital Communication and Coding*, New York, McGraw-Hill, 1979, pp. 40-42.

[9] R.G. Gallager, "Conflict resolution in random access broadcast networks", preliminary manuscript, 1979.

[10] J. Mosely, "An efficient contention resolution algorithm for multiple access channels", Tech. Rpt. LIDS-TH-918, Lab. Info. Dec. Sys., Mass. Inst. of Tech., Cambridge, MA, June 1979.

[11] N. Pippenger, "Bounds on the performance of protocols for a multi-access broadcast channel", Res. Rpt. 7742 (#33525), IBM Res. Ctr., Yorktown Heights, NY, June 28, 1979.

[12] P. Humblet, Private Communication, November 1979.

[13] M. Molle, "On the capacity of infinite population multiple access protocols", submitted to *IEEE Trans. Info. Th.*, April 1980.

[14] L. Kleinrock and F.A. Tobagi, "Packet switching in radio channels: Parity I – Carrier sense multiple-access modes and their throughput-delay characteristics", *IEEE Trans. Comm.*, vol. COM-23, pp. 1400-1416, December 1975.

[15] L.G. Roberts, "Dynamic allocation of satellite capacity through packet reservations", *Proc. Nat. Comp. Conf.*, pp. 711-716, 1973.

[16] J.F. Hayes, "An adaptive technique for local distribution", *IEEE Trans. Comm.*, vol. COM-26, pp. 1178-1186, August 1978.

REFERENCES

SPREAD-SPECTRUM MULTIPLE-ACCESS COMMUNICATIONS

Michael B. Pursley
Coordinated Science Laboratory
University of Illinois
Urbana, Illinois 61801 USA

In a direct-sequence spread-spectrum multiple-access communications system several asynchronous signals simultaneously occupy the same channel. Each of the signals employs a signature sequence which is selected to have certain desirable correlation properties. For multiple-access communications the primary goal is to be able to separate the spread-spectrum signals at the receiver even though they occupy the same bandwidth at the same time. This problem is considered in the sections which follow for various forms of direct-sequence spread-spectrum modulation including binary phase-shift keying, quadriphase-shift keying, and minimum-shift keying. The emphasis is on the analysis of system performance rather than on the selection of signature sequences. Hence this material complements the recent paper of Sarwate and Pursley (1980) which examines in detail the problem of signature sequence selection.

In Section I we give a random signal model of direct-sequence spread-spectrum modulation which is employed to illustrate the basic concepts of spread-spectrum multiple-access (SSMA) communications. In Section II binary direct-sequence spread-spectrum is considered and the multiple-access interference is characterized in terms of the continuous-time partial crosscorrelation functions for the direct-sequence code waveforms. This multiple-access interference is analyzed in Section III where maximum and mean-squared values of the multiple-access interference are related to maximum and mean-squared values of the continuous-time crosscorrelation functions. Several important properties of the continuous-time partial crosscorrelation functions are given in Section IV including the properties that are useful in the evaluation of the maximum and mean-squared values of the multiple-access interference. The analysis of Section IV is carried out for an arbitrary time-limited chip waveform, so that it can be applied to quite general systems such as quadriphase-shift-keyed and minimum-shift-keyed spread-spectrum communications systems. The signal-to-noise ratio for general quaternary direct-sequence spread-spectrum multiple-access communications is the main topic for Section V. Quadriphase-shift-keying and minimum-shift-keying are important special cases of the general quaternary direct-sequence spread-spectrum modulation that is analyzed in this section. Both quadriphase-shift-keying and minimum-shift-keying produce constant envelope signals. Numerical values of the signal-to-noise ratio are given for these two types of spread-spectrum modulation.

I. SPREAD-SPECTRUM COMMUNICATIONS: A RANDOM SIGNAL MODEL

In order to illustrate the basic concepts of spread-spectrum multiple-access (SSMA) communications, we first consider a system in which the spectral spreading signals are random processes. In the analysis of this system, we make no attempt to be mathematically precise because such spectral spreading signals are not of interest for system implementations. Instead, the analysis presented in this section is for the purpose of developing a physical insight into the basic concepts of SSMA communications systems. In later sections, more precise analyses will be presented after we first introduce a more specific system model which is implementable.

Let $V(t)$ be a zero-mean stationary Gaussian random process which has autocorrelation function $R_V(\tau)$ given by

$$R_V(\tau) = \begin{cases} (T_c)^{-1}(T_c - |\tau|), & |\tau| \leq T_c \\ 0, & |\tau| > T_c \,. \end{cases}$$

The actual shape of the autocorrelation function is not particularly important; all that is really required is that $R_V(\tau) \approx 0$ for $\tau > T_c$. The spectral density for the random process $V(t)$ with the above autocorrelation function is

$$S_V(f) = T_c[\mathrm{sinc}(fT_c)]^2$$

where $\mathrm{sinc}(x) = (\pi x)^{-1}\sin(\pi x)$. Define the bandwidth B of the process $V(t)$ to be the first zero-crossing of its spectrum; that is, B is the smallest positive value of f for which $S_V(f) = 0$. Hence, $B = (T_c)^{-1}$.

Let s(t) be a binary data signal which consists of a sequence of rectangular pulses, each of which is of duration T and amplitude $\pm$ A, where A is positive. For t in the range $0 \leq t < T$, either $s(t) = u_0(t) \triangleq Ap_T(t)$ or $s(t) = u_1(t) \triangleq -Ap_T(t)$, where $p_T(t)$ is the unit-amplitude rectangular pulse which starts at t = 0 and is of duration T. It is assumed that $T \gg T_c$ which implies that $B \gg T^{-1}$. Consequently, the bandwidth of V(t) is much greater than the bandwidth of the data signal.

For simplicity, the channel model that will be employed in this section is a noiseless baseband channel, which is sufficient to illustrate the key concepts. In subsequent sections, more complicated (and more physically realistic) channel models will be considered. The signal s(t)V(t) is transmitted over the baseband channel to a receiver which is to decide whether s(t) = A or s(t) = -A during a given interval. The baseband receiver is a correlation receiver which multiplies the received signal s(t)V(t) by V(t) and then integrates the product. For the data pulse transmitted during the interval [0,T) the decision statistic is

$$Z = \int_0^T s(t)V^2(t)dt. \tag{1.1}$$

If $s(t) = u_m(t)$ for $0 \leq t < T$ then

$$Z = Z^{(m)} = (-1)^m A \int_0^T V^2(t)dt \tag{1.2}$$

for m = 0 or 1. Ergodicity implies that with high probability

$$T^{-1} \int_0^T V^2(t)dt \approx E\{V^2(t)\} = R_V(0) = 1 \tag{1.3}$$

since $T >> T_c$. As a result

$$Z \approx (-1)^m AT \quad (1.4)$$

Thus, in the absence of any channel noise, the transmitted information is recovered from $s(t)V(t)$. If $Z \geq 0$ then the optimum receiver decides a positive pulse was sent ($m = 0$), if $Z < 0$ the receiver decides a negative pulse was sent ($m = 1$). Notice however that the receiver must know $V(t)$ exactly, which is the reason that this conceptual model is not of interest for practical implementations. In practice the characteristics of $V(t)$ are approximated by a "pseudo-noise" waveform which is usually derived from a pseudorandom sequence. The result (1.4) does not take into account the possibility of any interfering signals such as from other authorized transmitters in a multiple-access system, from unauthorized transmitters or authorized transmitters which are not part of the system, or from multipath propagation. The effects of such interfering signals are considered in what follows.

(i) Multiple-Access Capability. Suppose that there are K transmitted signals, each of the form described above. Each transmitter is assigned a different process $V_k(t)$, $1 \leq k \leq K$, each of which is a zero-mean stationary Gaussian process with autocorrelation function $R_V(\tau)$. The processes are mutually independent. The received signal is now

$$r(t) = \sum_{k=1}^{K} s_k(t)V_k(t) .$$

The output of the i-th correlation receiver is given by

$$Z_i = \int_0^T r(t)V_i(t)dt$$

$$= \int_0^T s_i(t)V_i^2(t)dt + \sum_{\substack{k=1\\k\neq i}}^{K} \int_0^T s_k(t)V_k(t)V_i(t)dt .$$

If $s_i(t) = u_m(t)$ for $0 \leq t < T$ then

$$Z_i = Z_i^{(m)} \approx (-1)^m AT + \sum_{k\neq i} \alpha_k A \int_0^T V_k(t)V_i(t)dt$$

where α_k is +1 or -1 depending on whether $s_k(t)$ is $u_0(t)$ or $u_1(t)$. Since $V_k(t)$ and $V_i(t)$ are independent zero-mean processes for $k\neq i$,

$$T^{-1}\int_0^T V_k(t)V_i(t)dt \approx 0 \tag{1.5}$$

so that

$$Z_i^{(m)} \approx (-1)^m AT \tag{1.6}$$

which is the same as (1.4). Thus, to within the accuracy of the approximation (1.5), the other signals in the system do not interfere with the transmission of the i-th signal. This demonstrates the multiple-access capability of spread spectrum.

(ii) Anti-Interference Capability. Suppose that the received signal is

$$r(t) = s(t)V(t) + A' ; \tag{1.7}$$

that is, a dc signal has been added to the desired signal. The dc signal level A' can be either positive or negative. The correlation receiver output for $s(t) = u_m(t)$ is

$$Z^{(m)} = \int_0^T s(t)V^2(t)dt + A' \int_0^T V(t)dt$$

$$\approx (-1)^m AT + A'T\{T^{-1} \int_0^T V(t)dt\} . \tag{1.8}$$

The interference is negligible provided

$$\frac{A'}{A} \{\frac{1}{T} \int_0^T V(t)dt\} \approx 0, \tag{1.9}$$

which is true if $A' \approx A$ since $V(t)$ has zero mean and T is large relative to T_c. This demonstrates the dc interference rejection capability of a spread-spectrum system. The ac interference rejection can be demonstrated via a similar argument which makes use of the fact that

$$\frac{1}{T} \int_0^T V(t)\cos \omega_0 t \; dt \approx 0 . \tag{1.10}$$

(iii) Anti-Multipath Capability. In a simplified baseband model of a multipath channel, the received signal might be

$$r(t) = s(t)V(t) - \beta s(t-\tau)V(t-\tau) \tag{1.11}$$

where β is in the range $0 \leq \beta < 1$. If, for instance, $s(t) = A$ for all t and $\tau > T_c$ then the output of the correlation receiver is

$$Z \approx AT - \beta A \int_0^T V(t)V(t-\tau)dt$$

$$\approx AT\{1 - \beta \, R_V(\tau)\} \approx AT. \tag{1.12}$$

Again, to within the accuracy of the approximation employed, the multipath signal does not interfere with the output of the correlation receiver provided that $\tau > T_c$.

In all these cases discussed above, the key feature is that the interfering signal has small correlation with V(t). As a result, the interfering signal produces only a negligible change in the output of the correlation receiver. The integrator tends to "smooth" the interfering signals, but it produces a large peak in response to the desired signal since it is matched to the rectangular data pulse.

A corresponding explanation can be given in terms of the frequency domain. The receiver "strips off" the spectral spreading signal from the data pulse and then passes the pulse through a narrowband (relative to B) filter. However, the interfering signals remain (or become) wideband when multiplied by V(t), so relatively little energy is passed by the narrow-band filter.

II. INTRODUCTION TO BINARY DIRECT-SEQUENCE SSMA COMMUNICATIONS

In the binary direct-sequence form of spread-spectrum modulation, a baseband signal of the form

$$x(t) = \sum_{j=-\infty}^{\infty} x_j \psi(t-jT_c) \tag{2.1}$$

is employed as the spectral-spreading signal. In (2.1) the sequence (x_j) is a periodic sequence of elements of $\{+1,-1\}$, and ψ is a time-limited signal (limited to $[0,T_c]$) for which

$$T_c^{-1} \int_0^{T_c} \psi^2(t)dt = 1. \tag{2.2}$$

The most common choice for the signal ψ is

$$\psi(t) = p_{T_c}(t) \triangleq \begin{cases} 1, & 0 \leq t < T_c \\ 0, & \text{otherwise} \end{cases} .$$

This is the rectangular pulse of duration T_c which starts at $t = 0$. Another signal of interest is the sine pulse

$$\psi(t) = \sqrt{2}\,\sin(\pi t/T_c)p_{T_c}(t) .$$

The rectangular pulse is employed in a phase-shift-key (PSK) system, and the sine pulse is the basic waveform for a minimum-shift-key (MSK) system. It is common to refer to $\psi(t)$, $0 \leq t \leq T_c$, as the <u>chip waveform</u>.

The binary data signal $b(t)$ is given by

$$b(t) = \sum_{\ell=-\infty}^{\infty} b_\ell p_T(t-\ell T) \tag{2.3}$$

where $p_T(t)$ is the rectangular pulse of duration T which starts at $t = 0$ and $b = (b_\ell)$ is the binary data sequence (i.e., $b_\ell \in \{+1,-1\}$ for each ℓ). The baseband spread-spectrum signal is then $v(t) = x(t)b(t)$. The sequence $x = (x_j)$, which is called the signature sequence, is a periodic sequence which satisfies $x_j = x_{j+N}$ for each j. That is, N is an integer multiple of the period of the signature sequence. The data pulse duration T is given by $T = NT_c$; therefore, the bandwidth of $v(t)$ is on the order of N times the bandwidth of $b(t)$.

The actual transmitted signal in a binary direct-sequence spread-spectrum system is

$$\begin{aligned} s(t) &= Av(t)\cos(\omega_c t + \theta) \\ &= Ax(t)b(t)\cos(\omega_c t + \theta) \end{aligned} \tag{2.4}$$

where ω_c is the carrier frequency and θ is an arbitrary phase angle. In a spread-spectrum multiple-access (SSMA) communications system there are K such signals which are simultaneously transmitted. For binary direct-sequence SSMA the signals are given by

$$s_k(t) = A\, a_k(t) b_k(t) \cos(\omega_c t + \theta_k) \tag{2.5}$$

for $1 \leq k \leq K$. The signal $a_k(t)$ is of the form (2.1); that is

$$a_k(t) = \sum_{j=-\infty}^{\infty} a_j^{(k)} \psi(t - jT_c) \tag{2.6}$$

where we have denoted the k-th signature sequence by $a^{(k)} = (a_j^{(k)})$. The k-th data signal $b_k(t)$ is of the form (2.3) with the k-th data sequence denoted by $b^{(k)} = (b_\ell^{(k)})$. In general, the phase angles θ_k, $1 \leq k \leq K$, are not the same because in practice the transmitters are not phase synchronous. Furthermore, the transmitters are not time synchronous, and the propagation delays for the various signals need not be the same. Thus, as illustrated by Figure 1, the received signal is given by

$$r(t) = n(t) + \sum_{k=1}^{K} s_k(t - \tau_k) \tag{2.7}$$

where n(t) is additive white Gaussian noise (thermal noise) with spectral density $\frac{1}{2}N_0$, and τ_k is the time delay associated with the k-th signal. From (2.5) and (2.7) we have

$$r(t) = n(t) + \sum_{k=1}^{K} A\, a_k(t - \tau_k) b_k(t - \tau_k) \cos(\omega_c t + \varphi_k) \tag{2.8}$$

where $\varphi_k = \theta_k - \omega_c \tau_k$. For our purposes only the <u>relative</u> time delays and phase angles need be considered, and so we assume $\varphi_i = 0$ and $\tau_i = 0$ in the analysis of the receiver which is matched to the i-th signal. In

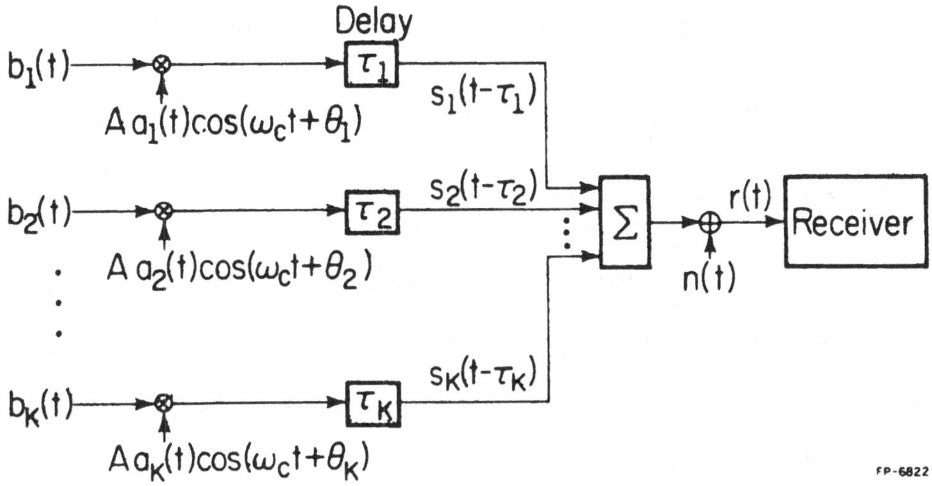

Figure 1. Binary Direct-Sequence SSMA Communications System Model

addition, there is no loss in generality in assuming that $0 \leq \varphi_k < 2\pi$ and $0 \leq \tau_k < T$, $1 \leq k \leq K$, since we are only concerned with time delays modulo T and phase shifts modulo 2π.

The i-th receiver is assumed to be a correlation receiver (or matched filter) which is matched to the i-th signal. Thus the output of the i-th receiver is

$$Z_i = \int_0^T r(t) a_i(t) \cos \omega_c t \, dt \tag{2.9}$$

since $\varphi_i = \tau_i = 0$. From (2.7) and (2.9) it is seen that

$$Z_i = \eta_i + \sum_{k=1}^{K} \int_0^T s_k(t - \tau_k) a_i(t) \cos \omega_c t \, dt \tag{2.10}$$

where η_i is the random variable

$$\eta_i = \int_0^T n(t) a_i(t) \cos \omega_c t \, dt \; . \tag{2.11}$$

If $\omega_c \gg T_c^{-1}$ then a practical spread-spectrum receiver is such that the double frequency components of the integrand of (2.9) can be ignored. In all that follows, it is assumed that such components are negligible. Under this assumption, (2.10) and (2.5) imply

$$Z_i = \eta_i + \tfrac{1}{2}A \int_0^T b_i(t) dt + \sum_{\substack{k=1 \\ k \neq i}}^{K} \tfrac{1}{2}A [f_{k,i}(\tau_k) + \hat{f}_{k,i}(\tau_k)] \cos \varphi_k \tag{2.12}$$

where the functions $f_{k,i}$ and $\hat{f}_{k,i}$ are defined by

$$f_{k,i}(\tau) = \int_0^\tau b_k(t - \tau) a_k(t - \tau) a_i(t) dt \tag{2.13}$$

and

$$\hat{f}_{k,i}(\tau) = \int_{\tau}^{T} b_k(t-\tau)a_k(t-\tau)a_i(t)dt \ . \tag{2.14}$$

Since $b_k(t)$ is a signal of the form given in (2.3) then $b_k(t-\tau) = b_{-1}^{(k)}$ for $0 \le t < \tau$. Thus (2.13) implies that for $0 \le \tau \le T$

$$f_{k,i}(\tau) = b_{-1}^{(k)} \int_{0}^{\tau} a_k(t-\tau)a_i(t)dt \ . \tag{2.15}$$

Similarly, $b_k(t-\tau) = b_0^{(k)}$ for $\tau \le t \le T$, and therefore (2.14) implies

$$\hat{f}_{k,i}(\tau) = b_0^{(k)} \int_{\tau}^{T} a_k(t-\tau)a_i(t)dt \ . \tag{2.16}$$

In view of (2.12), (2.15), and (2.16) it is of interest to consider the functions

$$R_{k,i}(\tau) = \int_{0}^{\tau} a_k(t-\tau)a_i(t)dt, \ 0 \le \tau \le T, \tag{2.17}$$

and

$$\hat{R}_{k,i}(\tau) = \int_{\tau}^{T} a_k(t-\tau)a_i(t)dt, \ 0 \le \tau \le T, \tag{2.18}$$

which are known as the continuous-time partial crosscorrelation functions. These functions were introduced in [Pursley (1974)] and they have been employed in the analysis of binary direct-sequence SSMA communications in several subsequent papers including [Pursley (1977)] and [Borth and Pursley (1979)]. Their effect on the receiver output is via the quantity

$$I_{k,i}(\underline{b}_k,\tau,\varphi) = T^{-1}[b_{-1}^{(k)}R_{k,i}(\tau) + b_0^{(k)}\hat{R}_{k,i}(\tau)]\cos\varphi \tag{2.19}$$

which we refer to as the (normalized) multiple-access interference at the output of the i-th receiver due to the k-th signal. In (2.19) the vector $\underline{b}_k$ is the vector $(b_{-1}^{(k)}, b_0^{(k)})$ of two consecutive data bits from the k-th data source.

The expression for Z_i given in (2.12) can now be written as

$$Z_i = \eta_i + \tfrac{1}{2}AT\{b_0^{(i)} + \sum_{k\neq i} I_{k,i}(\underline{b}_k,\tau_k,\varphi_k)\} \tag{2.20}$$

where $\sum_{k\neq i}$ denotes the sum over all integers k such that $k\neq i$ and $1 \leq k \leq K$. The first term on the right-hand side of (2.20) is due to the channel noise, the term $\tfrac{1}{2}AT\, b_0^{(i)}$ is due to the i-th signal, and the final term is the multiple-access interference due to the K-1 signals $\{s_k(t)\colon 1 \leq k \leq K,\ k\neq i\}$. If K = 1 the system is not a multiple-access system, the last term is not present, and the analysis of the system is straightforward. The difficulty that arises in a direct-sequence spread-spectrum multiple-access system is due to the multiple-access interference component of the received signal. Thus, we are mainly interested in the quantity

$$\mathcal{I}_i(\underline{b},\underline{\tau},\underline{\varphi}) = \sum_{k\neq i} I_{k,i}(\underline{b}_k,\tau_k,\varphi_k), \tag{2.21}$$

where

$$\underline{b} = (b_{-1}^{(1)},b_0^{(1)},b_{-1}^{(2)},b_0^{(2)},\ldots,b_{-1}^{(K)},b_0^{(K)})$$

is a vector of data symbols ($b_\ell^{(k)} \in \{+1,-1\}$),

$$\underline{\tau} = (\tau_1,\tau_2,\ldots,\tau_K)$$

is the vector of time delays, and

$$\underline{\varphi} = (\varphi_1,\varphi_2,\ldots,\varphi_K)$$

is the vector of phase angles. In view of the definition (2.21), the receiver output can be written as

$$Z_i = \eta_i + \tfrac{1}{2}AT\{b_0^{(i)} + \mathcal{I}_i(\underline{b},\underline{\tau},\underline{\varphi})\}. \tag{2.22}$$

The decision made by the receiver as to which pulse was transmitted (i.e., $b_0^{(i)} = +1$ or -1) is based on the observation Z_i. If $Z_i \geq 0$ then the receiver decides a positive pulse was sent (i.e., $b_0^{(i)} = +1$). Otherwise, the decision is that a negative pulse was sent (i.e., $b_0^{(i)} = -1$). If the decision is to be based solely on the observation Z_i, then the above decision rule is the best possible. Furthermore, Z_i is a sufficient statistic if $K = 1$. If $K > 1$ a more complex receiver may provide a statistic which permits a more nearly optimal decision. However, since correlation receivers (or matched filters) are relatively simple to implement, the vast majority of direct-sequence spread-spectrum systems employ correlation receivers even though they may be suboptimal in a multiple-access environment. This is apparently due to the relatively small performance improvement that can be obtained with the relatively complex optimum receiver. Consequently, our presentation will be confined to performance analysis for direct-sequence spread-spectrum multiple-access communications systems with correlation receivers.

One of the performance measures that will be considered is the probability of error. From (2.22) and the fact that η_i is a zero-mean Gaussian random variable with variance $\tfrac{1}{4}N_0T$, it follows that the conditional probability of error given that a positive pulse is transmitted (i.e., given $b_0^{(i)} = +1$) is expressed in terms of the standard Gaussian distribution function Φ by

$$P_{e,i}(\underline{b},\underline{\tau},\underline{\varphi}) = P[Z_i < 0 | b_0^{(i)} = +1]$$
$$= P[(2\eta_i/AT) < -1 - \mathcal{I}_i(\underline{b},\underline{\tau},\underline{\varphi})]$$
$$= \Phi(\sigma^{-1}[-1 - \mathcal{I}_i(\underline{b},\underline{\tau},\underline{\varphi})])$$
$$= Q(\sigma^{-1}[1 + \mathcal{I}_i(\underline{b},\underline{\tau},\underline{\varphi})]) \qquad (2.23)$$

for each $\underline{\tau}$ and $\underline{\varphi}$ and for each $\underline{b}$ such that $b_0^{(i)} = +1$. In (2.23) the function Q is defined by

$$Q(x) = 1 - \Phi(x) = \int_x^\infty (2\pi)^{-\frac{1}{2}} e^{-y^2/2}\, dy$$

and the quantity σ is given by

$$\sigma = 2(AT)^{-1} \{\mathrm{Var}\ \eta_i\}^{\frac{1}{2}} = (2\mathcal{E}_b/N_0)^{-\frac{1}{2}} \qquad (2.24)$$

where $\mathcal{E}_b$ is the energy per data bit ($\mathcal{E}_b = \frac{1}{2}A^2T$).

III. MULTIPLE-ACCESS INTERFERENCE

From (2.20) it is clear that in order to analyze the receiver output Z_i for a binary direct-sequence SSMA system, it is necessary to analyze the multiple-access interference $I_{k,i}(\underline{b}_k,\tau,\varphi)$ for $0 \leq \tau \leq T$, $0 \leq \varphi \leq 2\pi$, $b_\ell^{(k)} \in \{+1,-1\}$, and $k \neq i$. In particular, in order to determine the maximum probability of error it is necessary to determine the maximum value of the magnitude of the multiple-access interference. The expression for the probability of error in (2.23) implies that for $b_0^{(i)} = +1$ the maximum value of $P_{e,i}(\underline{b},\underline{\tau},\underline{\varphi})$ corresponds to the minimum value of $\mathcal{I}_i(\underline{b},\underline{\tau},\underline{\varphi})$. From (2.21) and (2.19) it is seen that

$$\min\{\mathcal{I}_i(\underline{b},\underline{\tau},\underline{\varphi})\} = -\max\{\mathcal{I}_i(\underline{b},\underline{\tau},\underline{\varphi})\}$$
$$= -\max\{|\mathcal{I}_i(\underline{b},\underline{\tau},\underline{\varphi})|\}$$
$$= -\sum_{k\neq i} \max\{|I_{k,i}(\underline{b}_k,\tau_k,\varphi_k)|\} \tag{3.1}$$

where the minimization and maximization are with respect to $\underline{b}$, $\underline{\tau}$, and $\underline{\varphi}$.

At this point it is helpful to introduce the continuous-time periodic crosscorrelation functions $\mathcal{R}_{k,i}(\tau)$, $1 \leq k \leq K$ and $1 \leq i \leq K$, which are defined by

$$\mathcal{R}_{k,i}(\tau) = \int_0^T a_k(t)a_i(t+\tau)dt \tag{3.2}$$

for all τ, $-\infty < \tau < \infty$. It is assumed that for each k, the period of $a_k(t)$ is a divisor of T. This is equivalent to the assumption that $N = T/T_c$ is a multiple of the period of the sequence $\underline{a}^{(k)} = (a_j^{(k)})$, and it implies that for each τ, $\mathcal{R}_{k,i}(\tau) = \mathcal{R}_{k,i}(\tau+T)$.

The functions $\mathcal{R}_{k,i}$, $R_{k,i}$, and $\hat{R}_{k,i}$ are related in the sense that any two of them determine the third. Specifically, from (2.17) and (2.18) we see that for τ in the range $0 \leq \tau \leq T$,

$$\hat{R}_{k,i}(\tau) + R_{k,i}(\tau) = \int_0^T a_k(t-\tau)a_i(t)dt$$
$$= \int_{-\tau}^{T-\tau} a_k(u)a_i(u+\tau)du$$
$$= \int_0^T a_k(u)a_i(u+\tau)du \quad .$$

Therefore,

$$\mathcal{R}_{k,i}(\tau) = \hat{R}_{k,i}(\tau) + R_{k,i}(\tau), \quad 0 \le \tau \le T. \tag{3.3}$$

The primary application of (3.3) to the analysis of direct-sequence spread-spectrum multiple-access communications stems from the observation that if $b_{-1}^{(k)} = b_0^{(k)}$, then according to (2.19) the multiple-access interference satisfies

$$I_{k,i}(\underline{b}_k, \tau, \varphi) = [T^{-1} b_0^{(k)} \cos \varphi] \mathcal{R}_{k,i}(\tau) \tag{3.4}$$

for each $\tau \in [0,T]$ and each $\varphi \in [0,2\pi]$. Notice that for each τ and φ

$$|I_{k,i}(\underline{b}_k, \tau, \varphi)| \le |I_{k,i}(\underline{b}_k, \tau, 0)| = T^{-1} |\mathcal{R}_{k,i}(\tau)| , \tag{3.5}$$

and that this upper bound is achieved for $\varphi = 0$.

The above results are all for the case $b_{-1}^{(k)} = b_0^{(k)}$. If $b_{-1}^{(k)} \ne b_0^{(k)}$ then (2.19) implies

$$I_{k,i}(\underline{b}_k, \tau, \varphi) = [T^{-1} b_0^{(k)} \cos \varphi] \{\hat{R}_{k,i}(\tau) - R_{k,i}(\tau)\} . \tag{3.6}$$

Because of this fact it is convenient to define

$$\hat{\mathcal{R}}_{k,i}(\tau) = \hat{R}_{k,i}(\tau) - R_{k,i}(\tau), \quad 0 \le \tau \le T. \tag{3.7}$$

We refer to $\hat{\mathcal{R}}_{k,i}(\tau)$ as the <u>continuous-time odd crosscorrelation function</u> since it has the property

$$\hat{\mathcal{R}}_{k,i}(\tau) = -\hat{\mathcal{R}}_{i,k}(T - \tau), \tag{3.8}$$

whereas the periodic (or even) crosscorrelation function has the property

$$\mathcal{R}_{k,i}(\tau) = \mathcal{R}_{i,k}(T - \tau) . \tag{3.9}$$

These two properties follow from (3.3), (3.7), and the two identities

$$\hat{R}_{k,i}(\tau) = R_{i,k}(T-\tau), \quad 0 \le \tau \le T, \tag{3.10}$$

and

$$R_{k,i}(\tau) = \hat{R}_{i,k}(T-\tau), \quad 0 \le \tau \le T. \tag{3.11}$$

In the situation $b_0^{(k)} \neq b_{-1}^{(k)}$, the multiple-access interference is given by

$$I_{k,i}(\underline{b}_k,\tau,\varphi) = [T^{-1} b_0^{(k)} \cos\varphi]\, \hat{\mathcal{R}}_{k,i}(\tau)\ . \tag{3.12}$$

The bound on the magnitude of this interference for $b_0^{(k)} \neq b_{-1}^{(k)}$ is given by

$$|I_{k,i}(\underline{b}_k,\tau,\varphi)| \le |I_{k,i}(\underline{b}_k,\tau,0)| = T^{-1}|\hat{\mathcal{R}}_{k,i}(\tau)|, \tag{3.13}$$

which is analogous to (3.5).

As discussed above, the determination of the maximum probability of error requires knowledge of the maximum value of the magnitude of the multiple-access interference $I_{k,i}(\underline{b}_k,\tau,\varphi)$. The inequalities (3.5) and (3.13) give the maximum with respect to the phase angle φ for each value of $\underline{b}_k$ and τ. The maximum with respect to $\underline{b}_k$ for each τ and φ is given by

$$\max\{|I_{k,i}(\underline{b}_k,\tau,\varphi)|\} = T^{-1}|\cos\varphi|\ \max\{|\mathcal{R}_{k,i}(\tau)|,|\hat{\mathcal{R}}_{k,i}(\tau)|\} \tag{3.14}$$

$$= T^{-1}|\cos\varphi|\{|\hat{R}_{k,i}(\tau)| + |R_{k,i}(\tau)|\}\ . \tag{3.15}$$

The value of $\underline{b}_k$ that achieves the upper bound is determined as follows. If $R_{k,i}(\tau)$ and $\hat{R}_{k,i}(\tau)$ are of the same sign then $b_{-1}^{(k)} = b_0^{(k)}$. In this case $b_0^{(k)} = +1$ if $\mathcal{R}_{k,i}(\tau)\cos\varphi \ge 0$, and $b_0^{(k)} = -1$ otherwise. On the other hand if $R_{k,i}(\tau)$ and $\hat{R}_{k,i}(\tau)$ are not of the same sign then

$b_{-1}^{(k)} = -b_0^{(k)}$. In this case $b_0^{(k)} = +1$ if $\hat{R}_{k,i}(\tau) \cos \varphi \geq 0$, and $b_0^{(k)} = -1$ otherwise. From the above we can also deduce the value of $\underline{b}_k$ which achieves the lower bound; that is, the value of $\underline{b}_k$ for which

$$I_{k,i}(\underline{b}_k,\tau,\varphi) = -T^{-1}|\cos \varphi|\{|\hat{R}_{k,i}(\tau)| + |R_{k,i}(\tau)|\} . \tag{3.16}$$

Notice that

$$I_{k,i}(-\underline{b}_k,\tau,\varphi) = -I_{k,i}(\underline{b}_k,\tau,\varphi) , \tag{3.17}$$

so that the same arguments as given above apply with $\underline{b}_k$ replaced by $-\underline{b}_k$.

The bounding of the multiple-access interference with respect to the phase angle φ is even more straightforward. A necessary condition for either the upper or the lower bound to be achieved is $|\cos \varphi| = 1$ (i.e., $\varphi = 0$ or $\varphi = \pi$). We find that for each τ

$$-T^{-1}\{|\hat{R}_{k,i}(\tau)|+|R_{k,i}(\tau)|\} \leq I_{k,i}(\underline{b}_k,\tau,\varphi) \leq T^{-1}\{|\hat{R}_{k,i}(\tau)|+|R_{k,i}(\tau)|\}. \tag{3.18}$$

The bounding of the multiple-access interference with respect to the variable τ is a slightly more difficult problem. It is clear from results such as (3.14) and (3.15) that the quantity

$$\begin{aligned} \mathcal{R}_{max}(k,i) &= \max\{|\hat{R}_{k,i}(\tau)| + |R_{k,i}(\tau)| \ : 0 \leq \tau \leq T\} \\ &= \max\{\max\{|\mathcal{R}_{k,i}(\tau)|, |\hat{\mathcal{R}}_{k,i}(\tau)|\} : 0 \leq \tau \leq T\} \end{aligned} \tag{3.19}$$

must be known in order to determine the maximum interference. The problem of evaluating $\mathcal{R}_{max}(k,i)$ is considered in the next section.

As an alternative to the maximum interference, the mean-squared value of the multiple-access interference is of interest, and it leads to the consideration of the signal-to-noise ratio as a measure of performance. We assume that $\underline{b}$, $\underline{\tau}$, and $\underline{\varphi}$ are independent random vectors since they arise from unrelated physical phenomena. We further assume the random variables $b_\ell^{(k)}$ and $b_n^{(j)}$ are independent if $k \neq j$ or if $n \neq \ell$, τ_k and τ_j are independent if $k \neq j$, and φ_k and φ_j are independent if $k \neq j$.

Notice that if $\underline{\theta} = (\theta_1, \theta_2, \ldots, \theta_K)$ is the vector of phase angles for the signals $s_k(t)$ defined by (2.5), then $\underline{\varphi} = \underline{\theta} - \omega_c \underline{\tau}$. Recall that all phase angles are considered modulo 2π. The fact that $\theta_k \pmod{2\pi}$ is uniform on $[0,2\pi]$ implies that for each <u>constant</u> τ, $\varphi \triangleq \theta_k - \omega_c \tau \pmod{2\pi}$ is uniform on $[0,2\pi]$. It follows that if τ_k and θ_k are independent random variables then the conditional density of $\varphi_k = \theta_k - \omega_c \tau_k \pmod{2\pi}$ given τ_k is a uniform density on $[0,2\pi]$. Since this implies the marginal density of φ_k is also uniform on $[0,2\pi]$, then φ_k and τ_k must be independent. Notice that these arguments impose no restrictions on the distribution of the random variable τ_k. As an example consider the situation where $\omega_c T = 2\pi n$ for some positive integer n. If τ_k is uniform on $[0,T]$ then both $\omega_c \tau_k \pmod{2\pi}$ and $-\omega_c \tau_k \pmod{2\pi}$ are uniform on $[0,2\pi]$. Therefore, if Θ and Θ' are random variables which take values in $[0,2\pi]$ and are defined by $\Theta = \theta_k \pmod{2\pi}$ and $\Theta' = -\omega_c \tau_k \pmod{2\pi}$ then Θ and Θ' are independent and identically distributed (each is uniform on $[0,2\pi]$). If $\Theta'' = \Theta + \Theta'$ then Θ'' takes values in $[0,4\pi]$ and has a triangular density function. However, if $\hat{\Theta}$ is a random variable which takes values in $[0,2\pi]$ and satisfies $\hat{\Theta} = \Theta'' \pmod{2\pi}$ then $\hat{\Theta}$ is uniform on $[0,2\pi]$. That is, although the sum of two independent random variables each of which is uniform on $[0,2\pi]$ is a random variable with a triangular density, the modulo-2π sum of two such random variables is a random variable which is also uniform on $[0,2\pi]$. It is easy to show that in this example $\hat{\Theta}$ and Θ' are independent. In fact $\{\Theta,\Theta',\hat{\Theta}\}$ are pairwise independent (but obviously they are not mutually independent since any pair determines the third).

From (2.19) we see that the independence assumption implies that

$$E\{I_{k,i}(\underline{b}_k,\tau_k,\varphi_k)\} = 0 \tag{3.20}$$

since φ_k is uniform on $[0,2\pi]$. If $P(b_{-1}^{(k)} = +1) = P(b_0^{(k)} = +1) = \frac{1}{2}$ then the variance of $I_{k,i}(\underline{b}_k,\tau_k,\varphi_k)$ is given by

$$\sigma_{k,i}^2 = \mathrm{Var}\{I_{k,i}(\underline{b}_k,\tau_k,\varphi_k)\} \tag{3.21}$$

$$= \tfrac{1}{2}\, T^{-2}\{\int_0^T T^{-1}[R_{k,i}^2(\tau) + \hat{R}_{k,i}^2(\tau)]d\tau\}, \tag{3.22}$$

since $E\{b_\ell^{(k)} b_j^{(k)}\}$ is 1 for $\ell = j$ and 0 for $\ell \neq j$. Define the quantities $\mathcal{M}_{k,i}$ and $\hat{\mathcal{M}}_{k,i}$ by

$$\mathcal{M}_{k,i} = \int_0^T R_{k,i}^2(\tau)d\tau \tag{3.23}$$

and

$$\hat{\mathcal{M}}_{k,i} = \int_0^T \hat{R}_{k,i}^2(\tau)d\tau \tag{3.24}$$

so that $T^{-1}\mathcal{M}_{k,i}$ and $T^{-1}\hat{\mathcal{M}}_{k,i}$ are the mean-squared values of $R_{k,i}(\tau)$ and $\hat{R}_{k,i}(\tau)$, respectively. In terms of these quantities, the variance of the multiple-access interference is

$$\sigma_{k,i}^2 = \tfrac{1}{2} T^{-3}(\mathcal{m}_{k,i} + \hat{\mathcal{m}}_{k,i}) . \tag{3.25}$$

In the next section, the quantities $\mathcal{m}_{k,i}$ and $\hat{\mathcal{m}}_{k,i}$ are examined more carefully and several results are given which are extremely useful in the evaluation of (3.25).

One reason that the quantity $\sigma_{k,i}^2$ is of interest stems from the consideration of the signal-to-noise ratio. As in [Pursley (1974), (1977)] we define the signal-to-noise ratio for the i-th receiver by

$$\mathrm{SNR}_i = E\{Z_i | b_0^{(i)} = +1\}[\mathrm{Var}\{Z_i | b_0^{(i)} = +1\}]^{-\frac{1}{2}} . \tag{3.26}$$

Notice that $I_{k,i}(\underline{b}_k,\tau_k,\varphi_k)$ does not depend on $b_0^{(i)}$, and so (2.20) and (3.20) imply that

$$E\{Z_i | b_0^{(i)} = +1\} = E\{\eta_i\} + \tfrac{1}{2} AT[1 + \sum_{k \neq i} E\{I_{k,i}(\underline{b}_k,\tau_k,\varphi_k)\}] = \tfrac{1}{2} AT. \tag{3.27}$$

In addition, $I_{k,i}(\underline{b}_k,\tau_k,\varphi_k)$ and $I_{j,i}(\underline{b}_j,\tau_j,\varphi_j)$ are independent for $k \neq j$ so

$$\begin{aligned}\mathrm{Var}\{Z_i | b_0^{(i)} = +1\} &= \mathrm{Var}\{\eta_i\} + (\tfrac{1}{2} AT)^2 \sum_{k \neq i} \mathrm{Var}\{I_{k,i}(\underline{b}_k,\tau_k,\varphi_k)\} \\ &= \tfrac{1}{4} N_0 T + \tfrac{1}{4} A^2 T^2 \sum_{k \neq i} \sigma_{k,i}^2 .\end{aligned} \tag{3.28}$$

From (3.26)-(3.28) we conclude that

$$\mathrm{SNR}_i = \left\{\frac{N_0}{A^2 T} + \sum_{k \neq i} \sigma_{k,i}^2\right\}^{-\frac{1}{2}} = \left\{\frac{N_0}{2\mathcal{E}_b} + \sum_{k \neq i} \sigma_{k,i}^2\right\}^{-\frac{1}{2}} . \tag{3.29}$$

For comparison with the results of [Pursley (1977)] we define

$$r_{k,i} = 6N^3 \sigma_{k,i}^2 \quad . \tag{3.30}$$

Several properties of the parameter $r_{k,i}$ are given in [Pursley and Sarwate (1977a,1977b)] for the special case in which the chip waveform is a rectangular pulse (i.e., $\psi(t) = p_{T_c}(t)$). In the next section, we will examine the parameter $r_{k,i}$ (or equivalently $\sigma_{k,i}^2$) for an arbitrary time-limited chip waveform $\psi(t)$.

Results such as (3.29) are easily extended to the situation in which the signals $s_k(t)$, $1 \leq k \leq K$, are not required to have the same power. This permits the analysis of the so-called "near-far" problem of direct-sequence SSMA systems. We simply rewrite (2.8) as

$$r(t) = n(t) + \sum_{k=1}^{K} A_k a_k(t-\tau_k) b_k(t-\tau_k) \cos(\omega_c t + \varphi_k) \quad . \tag{3.31}$$

With this change (2.20) becomes

$$Z_i = \eta_i + \tfrac{1}{2} A_i T \{ b_0^{(i)} + \sum_{k \neq i} \epsilon_{k,i}^{\frac{1}{2}} I_{k,i}(\underline{b}_k, \tau_k, \varphi_k) \} \tag{3.32}$$

where $\epsilon_{k,i} = (A_k/A_i)^2$. We can then define

$$I'_{k,i}(\underline{b}_k, \tau_k, \varphi_k) = \epsilon_{k,i}^{\frac{1}{2}} I_{k,i}(\underline{b}_k, \tau_k, \varphi_k)$$

so that (2.22) is valid if $I_{k,i}$ is replaced by $I'_{k,i}$ in (2.21). Similarly, $I_{k,i}$ is replaced by $I'_{k,i}$ in (3.1). The results which generalize (3.26)-(3.29) are as follows. The conditional mean is

$$E\{Z_i \mid b_0^{(i)} = +1\} = \tfrac{1}{2} A_i T \tag{3.33}$$

and the conditional variance is

$$\mathrm{Var}\{Z_i \mid b_0^{(i)} = +1\} = \tfrac{1}{4} N_0 T + \tfrac{1}{4} A_i^2 T^2 \sum_{k \neq i} \epsilon_{k,i} \sigma_{k,i}^2 \quad . \tag{3.34}$$

From (3.26), (3.33), and (3.34) we conclude that if $\mathcal{E}_i = \frac{1}{2} A_i^2 T$ then

$$\mathrm{SNR}_i = \left\{ \frac{N_0}{A_i^2 T} + \sum_{k \neq i} \epsilon_{k,i} \, \sigma_{k,i}^2 \right\}^{-\frac{1}{2}} = \left\{ \frac{N_0}{2\mathcal{E}_i} + \sum_{k \neq i} \epsilon_{k,i} \, \sigma_{k,i}^2 \right\}^{-\frac{1}{2}} . \tag{3.35}$$

The signal-to-noise ratio is a very useful measure of performance for direct-sequence SSMA communications. First, it is an easy quantity to evaluate since it can be computed from the aperiodic autocorrelation functions of the signature sequences (this is discussed further in Section IV). The numerical integrations and computations of cross-correlation functions that are required for the determination of the average probability of error are not needed in the evaluation of signal-to-noise ratio. Furthermore, as shown in recent papers [Yao (1977); Borth, Pursley, Sarwate, and Stark (1979); and Pursley, Sarwate and Stark (1980)], the signal-to-noise ratio can be used to obtain an accurate estimate of the average probability of error for most systems of interest. This approximation, which was suggested in [Pursley (1977)], is $P_{e,i} \approx 1 - \Phi(\mathrm{SNR}_i)$ where $P_{e,i}$ is the average probability of error for the i-th signal and Φ is the standard Gaussian distribution function. Results on the accuracy of this approximation can be obtained from the recent papers mentioned above.

IV. PROPERTIES OF THE CONTINUOUS-TIME CROSSCORRELATION FUNCTIONS

In the last section it was shown that the maximum multiple-access interference depends upon the quantity

$$\begin{aligned}\mathcal{R}_{max}(k,i) &= \max\{|R_{k,i}(\tau)|+|\hat{R}_{k,i}(\tau)| \ : \ 0 \le \tau \le T\} \\ &= \max\{\max\{|\mathcal{R}_{k,i}(\tau)|,|\hat{\mathcal{R}}_{k,i}(\tau)|\} \ : \ 0 \le \tau \le T\}. \end{aligned} \tag{4.1}$$

It was also shown that the variance of the multiple-access interference is given by

$$\sigma^2_{k,i} = \tfrac{1}{2} T^{-3} \int_0^T [R^2_{k,i}(\tau) + \hat{R}^2_{k,i}(\tau)]d\tau \ . \tag{4.2}$$

In this section we examine the continuous-time partial crosscorrelation functions $R_{k,i}(\tau)$ and $\hat{R}_{k,i}(\tau)$, and we give several important properties of these functions which are useful in the evaluation of the parameters $\mathcal{R}_{max}(k,i)$ and $\sigma^2_{k,i}$.

In order to obtain expressions for $R_{k,i}(\tau)$ and $\hat{R}_{k,i}(\tau)$ it is perhaps helpful to the reader to first consider the case $\tau = \ell T_c$ for some integer ℓ in the range $0 \le \ell \le N-1$. We find that

$$\begin{aligned} R_{k,i}(\ell T_c) &= \int_0^{\ell T_c} a_k(t-\ell T_c)a_i(t)dt \\ &= \sum_{j=0}^{\ell-1} a^{(k)}_{j-\ell}\, a^{(i)}_j \int_0^{T_c} \psi^2(t)dt \\ &= \left\{ \sum_{j=0}^{\ell-1} a^{(k)}_{j-\ell}\, a^{(i)}_j \right\} T_c \end{aligned} \tag{4.3}$$

and

$$\hat{R}_{k,i}(\ell T_c) = \int_{\ell T_c}^{T} a_k(t-\ell T_c)a_i(t)dt$$

$$= \sum_{j=\ell}^{N-1} a_{j-\ell}^{(k)} a_j^{(i)} \int_0^{T_c} \psi^2(t)dt$$

$$= \left\{ \sum_{j=0}^{N-1-\ell} a_j^{(k)} a_{j+\ell}^{(i)} \right\} T_c \; . \tag{4.4}$$

The quantities $R_{k,i}(\ell T_c)$ and $\hat{R}_{k,i}(\ell T_c)$ can be written in terms of the <u>aperiodic crosscorrelation function</u> $C_{k,i}(\cdot)$ which is defined by

$$C_{k,i}(\ell) = \begin{cases} \sum_{j=0}^{N-1-\ell} a_j^{(k)} a_{j+\ell}^{(i)} \; , & 0 \le \ell \le N-1 \quad (4.5a) \\ \sum_{j=0}^{N-1+\ell} a_{j-\ell}^{(k)} a_j^{(i)} \; , & 1-N \le \ell < 0, \quad (4.5b) \end{cases}$$

and $C_{k,i}(\ell) = 0$ for $|\ell| \ge N$. This function has been considered in [Pursley (1977), Pursley and Sarwate (1977a,1977b), and Sarwate and Pursley (1980)] where many of its important properties are given. Our interest here lies in the fact that (4.3) and (4.4) can be written as

$$R_{k,i}(\ell T_c) = C_{k,i}(\ell-N)T_c \tag{4.6}$$

and

$$\hat{R}_{k,i}(\ell T_c) = C_{k,i}(\ell)T_c \; . \tag{4.7}$$

As expected, $R_{k,i}(\tau)$ and $\hat{R}_{k,i}(\tau)$ do not depend on the chip waveform if τ is an integer multiple of the chip duration.

Now consider general values of τ in the range $0 \leq \tau < T$. For each such τ there is a unique integer ℓ in the range $0 \leq \ell \leq N-1$ which is such that $\ell T_c \leq \tau < (\ell+1)T_c$. The integer ℓ is given by $\ell = \lfloor \tau/T_c \rfloor$ where $\lfloor u \rfloor$ denotes the integer part of the real number u. From (2.6) and (2.18) it follows that for each τ in the range $0 \leq \tau < T$,

$$\hat{R}_{k,i}(\tau) = \sum_{j=0}^{N-1-\ell} a_j^{(k)} a_{j+\ell}^{(i)} \int_{\tau'}^{T_c} \psi(t)\psi(t-\tau')dt$$

$$+ \sum_{j=0}^{N-1-\ell'} a_j^{(k)} a_{j+\ell'}^{(i)} \int_0^{\tau'} \psi(t)\psi(t+T_c-\tau')dt \tag{4.8}$$

where $\ell = \lfloor \tau/T_c \rfloor$, $\ell' = \ell+1$, and $\tau' = \tau-\ell T_c$. Notice that each of the two sums which appear in the expression given in (4.8) has a sequence-dependent component and a component which depends only on the chip waveform. The sequence-dependent part can be expressed in terms of the aperiodic crosscorrelation function, and the waveform-dependent component can be written in terms of the partial autocorrelation functions for the chip waveform, which are defined by

$$\hat{R}_\psi(s) = \int_s^{T_c} \psi(t)\psi(t-s)dt\ , \quad 0 \leq s \leq T_c, \tag{4.9}$$

and

$$R_\psi(s) = \int_0^{s} \psi(t)\psi(t+T_c-s)dt\ , \quad 0 \leq s \leq T_c. \tag{4.10}$$

The dependence of $\hat{R}_{k,i}(\tau)$ on the sequences $a^{(k)}$ and $a^{(i)}$ and the chip waveform $\psi(t)$ is via the functions $C_{k,i}$, $\hat{R}_\psi$, and R_ψ. This is demonstrated by substitution from (4.5a), (4.9), and (4.10) into (4.8) which yields

and

$$R_\psi(s) = -s\cos(\pi s/T_c) + (T_c/\pi)\sin(\pi s/T_c) . \qquad (4.16)$$

It is convenient to define

$$c(\tau) = \cos(\pi\tau/T_c) , \qquad 0 \le \tau \le T_c , \qquad (4.17)$$

and

$$s(\tau) = \sin(\pi\tau/T_c) , \qquad 0 \le \tau \le T_c . \qquad (4.18)$$

By employing (4.15) and (4.16) in (4.11) and (4.12) we find that

$$\hat{R}_{k,i}(\tau) = (-1)^{\ell}\{[C_{k,i}(\ell) + C_{k,i}(\ell+1)]\left[\frac{T_c}{\pi}s(\tau) - (\tau-\ell T_c)c(\tau)\right] + C_{k,i}(\ell)T_c c(\tau)\} \qquad (4.19)$$

and

$$R_{k,i}(\tau) = (-1)^{\ell}\{[C_{k,i}(\ell-N) + C_{k,i}(\ell+1-N)]\left[\frac{T_c}{\pi}s(\tau) - (\tau-\ell T_c)c(\tau)\right] + C_{k,i}(\ell-N)T_c c(\tau)\} . \qquad (4.20)$$

For the general time-limited chip waveform the functions $\mathcal{R}_{k,i}$ and $\hat{\mathcal{R}}_{k,i}$, which are defined in (3.3) and (3.7), represent the interference between the k-th and i-th signals (see (3.4) and (3.12)). From (4.11) and (4.12) we see that

$$\mathcal{R}_{k,i}(\tau) = [C_{k,i}(\ell) + C_{k,i}(\ell-N)]\hat{R}_\psi(\tau-\ell T_c) + [C_{k,i}(\ell+1) + C_{k,i}(\ell+1-N)]R_\psi(\tau-\ell T_c) . \qquad (4.21)$$

The <u>periodic crosscorrelation function</u> for the sequences $a^{(k)}$ and $a^{(i)}$ is given by

$$\theta_{k,i}(\ell) = \sum_{j=0}^{N-1} a_j^{(k)} a_{j+\ell}^{(i)} \tag{4.22}$$

$$= C_{k,i}(\ell) + C_{k,i}(\ell-N) \tag{4.23}$$

for $0 \leq \ell \leq N-1$. This function is discussed in detail in [Sarwate and Pursley (1980)]. Because of (4.23), the continuous-time periodic crosscorrelation can be written as

$$\mathcal{R}_{k,i}(\tau) = \theta_{k,i}(\ell)\hat{R}_\psi(\tau-\ell T_c) + \theta_{k,i}(\ell+1)R_\psi(\tau-\ell T_c) \; . \tag{4.24}$$

Similarly, the continuous-time odd crosscorrelation function is given by

$$\hat{\mathcal{R}}_{k,i}(\tau) = [C_{k,i}(\ell) - C_{k,i}(\ell-N)]\hat{R}_\psi(\tau-\ell T_c)$$

$$+ [C_{k,i}(\ell+1) - C_{k,i}(\ell+1-N)]R_\psi(\tau-\ell T_c) \tag{4.25}$$

$$= \hat{\theta}_{k,i}(\ell)\hat{R}_\psi(\tau-\ell T_c) + \hat{\theta}_{k,i}(\ell+1)R_\psi(\tau-\ell T_c), \tag{4.26}$$

where $\hat{\theta}_{k,i}$ is the odd crosscorrelation function for the sequences $a^{(k)}$ and $a^{(i)}$, which is defined by

$$\hat{\theta}_{k,i}(\ell) = C_{k,i}(\ell) - C_{k,i}(\ell-N) \tag{4.27}$$

for $0 \leq \ell \leq N-1$. This function was first introduced by Massey and Uhran (1969). Several of its properties are presented in [Massey and Uhran (1975)], [Pursley and Roefs (1979)], and [Sarwate and Pursley (1980)].

From the discussion in Section III it is clear that the maximum value of the multiple-access interference depends on the quantity

$\mathcal{R}_{max}(k,i)$ which is defined in (3.19). In order to determine $\mathcal{R}_{max}(k,i)$ it is necessary to first determine

$$r_{max}(k,i) = \max\{|\mathcal{R}_{k,i}(\tau)| : 0 \le \tau \le T\} \tag{4.28}$$

and

$$\hat{r}_{max}(k,i) = \max\{|\hat{\mathcal{R}}_{k,i}(\tau)| : 0 \le \tau \le T\} . \tag{4.29}$$

The first step is to consider the following maximizations

$$r_\ell(k,i) = \max\{|\mathcal{R}_{k,i}(\tau)| : \ell T_c \le \tau \le (\ell+1)T_c\}$$

and

$$\hat{r}_\ell(k,i) = \max\{|\hat{\mathcal{R}}_{k,i}(\tau)| : \ell T_c \le \tau \le (\ell+1)T_c\} .$$

We will show that for an arbitrary chip waveform $\psi(t)$,

$$r_\ell(k,i) = T_c \max\{|\theta_{k,i}(\ell)|, |\theta_{k,i}(\ell+1)|\} \tag{4.30}$$

and

$$\hat{r}_\ell(k,i) = T_c \max\{|\hat{\theta}_{k,i}(\ell)|, |\hat{\theta}_{k,i}(\ell+1)|\} . \tag{4.31}$$

That is, the continuous-time periodic crosscorrelation functions have their maximum magnitudes at integer multiples of T_c.

In order to establish (4.30) we let ℓ be fixed and define $\alpha = \theta_{k,i}(\ell)$ and $\beta = \theta_{k,i}(\ell+1)$. For τ in the interval $[\ell T_c, (\ell+1)T_c]$ let $s = \tau - \ell T_c$ and define $\hat{y}(s) = \hat{R}_\psi(s)$ and $y(s) = R_\psi(s)$ for $0 \le s \le T_c$. Then from (4.24) we see that

$$|\mathcal{R}_{k,i}(\tau)| = |\alpha \hat{y}(s) + \beta y(s)| . \tag{4.32}$$

Similarly define $\hat{\alpha} = \hat{\theta}_{k,i}(\ell)$ and $\hat{\beta} = \hat{\theta}_{k,i}(\ell+1)$ so that (4.26) implies

$$|\hat{\mathcal{R}}_{k,i}(\tau)| = |\hat{\alpha} \hat{y}(s) + \hat{\beta} y(s)| . \tag{4.33}$$

Let $\mathcal{X}'$ be the set of all pairs $(\hat{y}(s),y(s))$ as s ranges over the interval $[0,T_c]$. We first show that

$$\mathcal{X}' \subset \mathcal{X} \triangleq \{(x_1,x_2): |x_1+x_2| \leq T_c, |x_1-x_2| \leq T_c\}. \tag{4.34}$$

Notice that (4.9) and (4.10) imply

$$|\hat{y}(s)|+|y(s)| \leq \int_s^{T_c} |\psi(t)\psi(t-s)|dt + \int_0^s |\psi(t)\psi(t+T_c-s)|dt$$

$$= \int_0^{T_c} |\tilde{\psi}(t)\tilde{\psi}(t-s)|dt \tag{4.35}$$

where

$$\tilde{\psi}(t) = \begin{cases} \psi(t), & 0 \leq t \leq T_c \\ \psi(t+T_c), & -T_c \leq t < 0. \end{cases}$$

From (4.35) we conclude that

$$|\hat{y}(s)|+|y(s)| \leq \int_0^{T_c} [\tilde{\psi}(t)]^2 dt = \int_0^{T_c} \psi^2(t)dt = T_c. \tag{4.36}$$

Next observe that $|\hat{y}(s) \pm y(s)| \leq |\hat{y}(s)| + |y(s)|$ and thus (4.36) implies (4.34). It is important to observe at this point that since $\hat{R}_\psi(0) = T_c$ and $R_\psi(0) = 0$ then the pair $(T_c,0) \in \mathcal{X}'$, and since $\hat{R}_\psi(T_c) = 0$ and $R_\psi(T_c) = T_c$ then the pair $(0,T_c) \in \mathcal{X}'$.

The function f defined on the convex set $\mathcal{X}$ by

$$f(x_1,x_2) = |\alpha x_1 + \beta x_2|, \quad (x_1,x_2) \in \mathcal{X}, \tag{4.37}$$

is a convex function. From (4.32) we see that

$$|\mathcal{R}_{k,i}(\tau)| = f(\hat{y}(s), y(s)),$$

and from (4.37) we see that $f(x_1,x_2) = f(-x_1,-x_2)$, and in particular

$$f(T_c,0) = f(-T_c,0),\ f(0,T_c) = f(0,-T_c). \tag{4.38}$$

Since f is convex, the maximum value of $f(x_1,x_2)$ is attained when (x_1,x_2) is an extreme point of the convex set $\mathcal{X}$. The extreme points of $\mathcal{X}$ are the four points $(0,T_c)$, $(T_c,0)$, $(0,-T_c)$, and $(-T_c,0)$. Thus (4.38) implies

$$\max\{f(x_1,x_2) : (x_1,x_2) \in \mathcal{X}\} = \max\{f(T_c,0), f(0,T_c)\} . \tag{4.39}$$

But since $(T_c,0)$ and $(0,T_c)$ are also in $\mathcal{X}'$ and $\mathcal{X}' \subset \mathcal{X}$ then

$$\max\{f(x_1,x_2) : (x_1,x_2) \in \mathcal{X}'\} = \max\{f(T_c,0), f(0,T_c)\} . \tag{4.40}$$

From (4.37) and (4.40) we conclude that

$$\max\{f(x_1,x_2) : (x_1,x_2) \in \mathcal{X}'\} = T_c \max\{|\alpha|, |\beta|\} \tag{4.41}$$

which is just (4.30). The above argument with $\alpha = \hat{\theta}_{k,i}(\ell)$ and $\beta = \hat{\theta}_{k,i}(\ell+1)$ establishes (4.31).

From the above we conclude that the peak crosscorrelations (both periodic and odd) occur when the relative delay τ is of the form $\tau = \ell T_c$ where ℓ is an integer. If the chip waveforms $\psi(t)$ are rectangular pulses then (4.24) and (4.25) imply

$$\mathcal{R}_{k,i}(\tau) = (T_c - s)\theta_{k,i}(\ell) + s\,\theta_{k,i}(\ell+1) \tag{4.42}$$

and

$$\hat{\mathcal{R}}_{k,i}(\tau) = (T_c - s)\hat{\theta}_{k,i}(\ell) + s\,\hat{\theta}_{k,i}(\ell+1) \tag{4.43}$$

where $s = \tau - \ell T_c$. Hence the conclusion that $|\hat{R}_{k,i}(\tau)|$ is maximized for $\tau = \ell T_c$ is just the conclusion that the right-hand side of (4.42) is maximized for $s = 0$ or $s = T_c$. The latter conclusion is trivial since the right-hand side of (4.42) is linear in s for $0 \le s \le T_c$. A similar observation can be made from (4.43) concerning the maximization of $|\hat{\hat{R}}_{k,i}(\tau)|$ in the special case of rectangular chip waveforms. However, the above results (i.e., (4.30) and (4.31)) substantiate the considerably less obvious conclusion that even for arbitrary time-limited chip waveforms, the crosscorrelation magnitudes are maximized for values of τ which are integer multiples of T_c. In particular, the above results imply that if

$$M_{k,i} = \max\{|\theta_{k,i}(\ell)| : \ell = 0,1,\ldots,N-1\} \tag{4.44}$$

and

$$\hat{M}_{k,i} = \max\{|\hat{\theta}_{k,i}(\ell)| : \ell = 0,1,\ldots,N-1\} \tag{4.45}$$

then (4.28) and (4.29) reduce to

$$r_{max}(k,i) = T_c\ M_{k,i} \tag{4.46}$$

and

$$\hat{r}_{max}(k,i) = T_c\ \hat{M}_{k,i}\ . \tag{4.47}$$

The parameters $M_{k,i}$ and $\hat{M}_{k,i}$ are discussed in detail in [Pursley and Roefs (1979), see equations (4) and (5) on page 1598] where numerical values are tabulated for several sets of sequences.

Returning now to the original problem posed in (4.1) we see that

$$R_{max}(k,i) = T_c \max\{M_{k,i}, \hat{M}_{k,i}\}\ , \tag{4.48}$$

values of $I_{k,i}(\underline{b}_k,\tau,\varphi)$ as large as $I_{max}(k,i)$. Hence, for many applications it is considerably more useful to consider average performance rather than worst-case performance. The principal measure of average performance that we consider is the signal-to-noise ratio which was defined in (3.26).

From (3.29) and (3.25) it is clear that the evaluation of the signal-to-noise ratio requires the evaluation of the sum of the parameters

$$\hat{\mathcal{M}}_{k,i} = \int_0^T \hat{R}_{k,i}^2(\tau)d\tau \tag{4.50}$$

and

$$\mathcal{M}_{k,i} = \int_0^T R_{k,i}^2(\tau)d\tau \ . \tag{4.51}$$

One fact of interest is that $\hat{\mathcal{M}}_{k,i} + \mathcal{M}_{k,i}$ can be computed from the continuous-time partial autocorrelation functions (which are defined by (2.17) and (2.18) with k = i). It follows from [Pursley and Sarwate (1977a), p. 513] that

$$\hat{\mathcal{M}}_{k,i} + \mathcal{M}_{k,i} = 2\int_0^T R_{k,k}(\tau)R_{i,i}(\tau)d\tau \ , \tag{4.52}$$

and therefore the signal-to-noise ratio depends only on the partial autocorrelation functions for the periodic signals $a_k(t)$, $1 \leq k \leq K$. The analysis could proceed from (4.52); however, we shall work with the individual parameters $\hat{\mathcal{M}}_{k,i}$ and $\mathcal{M}_{k,i}$ for the present. The final conclusions are the same for either approach.

From (4.11) we find that

$$\hat{\mathcal{m}}_{k,i} = \sum_{\ell=0}^{N-1} \int_{\ell T_c}^{(\ell+1)T_c} \hat{R}^2_{k,i}(\tau)d\tau = \sum_{\ell=0}^{N-1} \int_0^{T_c} \hat{R}^2_{k,i}(\tau+\ell T_c)d\tau$$

$$= \sum_{\ell=0}^{N-1} \{C^2_{k,i}(\ell)\,\hat{\mathcal{m}}_\psi + 2C_{k,i}(\ell)C_{k,i}(\ell+1)\,\mathcal{m}'_\psi + C^2_{k,i}(\ell+1)\,\mathcal{m}_\psi\} \quad (4.53)$$

where

$$\hat{\mathcal{m}}_\psi = \int_0^{T_c} \hat{R}^2_\psi(s)\,ds,\quad \mathcal{m}_\psi = \int_0^{T_c} R^2_\psi(s)\,ds, \quad (4.54)$$

and

$$\mathcal{m}'_\psi = \int_0^{T_c} R_\psi(s)\hat{R}_\psi(s)\,ds\ . \quad (4.55)$$

Of course since $\hat{R}_\psi(T_c-s) = R_\psi(s)$ then $\hat{\mathcal{m}}_\psi = \mathcal{m}_\psi$. The expression for $\mathcal{m}_{k,i}$ which is analogous to (4.53) is

$$\mathcal{m}_{k,i} = \sum_{\ell=-N}^{-1} \{C^2_{k,i}(\ell)\,\hat{\mathcal{m}}_\psi + 2C_{k,i}(\ell)C_{k,i}(\ell+1)\,\mathcal{m}'_\psi + C^2_{k,i}(\ell+1)\,\mathcal{m}_\psi\}\ , \quad (4.56)$$

which follows from (4.12). Combining (3.25), (4.53), and (4.56) we find

$$\sigma^2_{k,i} = T^{-3}\{\mu_{k,i}(0)\,\mathcal{m}_\psi + \mu_{k,i}(1)\,\mathcal{m}'_\psi\}, \quad (4.57)$$

where the parameter $\mu_{k,i}(n)$ is defined by

$$\mu_{k,i}(n) = \sum_{\ell=1-N}^{N-1} C_{k,i}(\ell)\ C_{k,i}(\ell+n)\ . \quad (4.58)$$

The importance of this parameter for spread-spectrum communications with a rectangular chip waveform has been known for some time [Pursley (1977)]. Fortunately, $\mu_{k,i}(n)$ is much easier to compute than might be inferred from (4.58). It is shown in [Pursley and Sarwate (1977b)] that

$$\mu_{k,i}(n) = \sum_{\ell=1-N}^{N-1} C_k(\ell)\, C_i(\ell+n) \tag{4.59}$$

where $C_k = C_{k,k}$ is the aperiodic autocorrelation function for the k-th signature sequence $(1 \leq k \leq K)$. Hence $\mu_{k,i}(n)$ can be computed from the K autocorrelation functions C_k, $1 \leq k \leq K$; the K^2 crosscorrelation functions $C_{k,i}$ are not required for the evaluation of $\mu_{k,i}(n)$.

For a rectangular waveform, $\psi(t) = p_{T_c}(t)$, we have previously observed that $\hat{R}_\psi(s) = T_c - s$ and $R_\psi(s) = s$. Consequently,

$$\mathcal{M}_\psi = \int_0^{T_c} s^2\, ds = T_c^3/3 , \tag{4.60}$$

$$\mathcal{M}'_\psi = \int_0^{T_c} s(T_c - s)ds = T_c^3/6 , \tag{4.61}$$

and thus (4.57) reduces to

$$\sigma_{k,i}^2 = (T_c/T)^3\, \{2\mu_{k,i}(0) + \mu_{k,i}(1)\}/6$$

$$= (6N^3)^{-1}\, \{2\mu_{k,i}(0) + \mu_{k,i}(1)\} , \tag{4.62}$$

which agrees with the results of [Pursley (1977)].

For the sine pulse, $\psi(t) = \sqrt{2}\,\sin(\pi t/T_c)p_{T_c}(t)$, we find that

$$\mathcal{M}_\psi = T_c^3\, (15 + 2\pi^2)/12\pi^2 \tag{4.63}$$

and

$$\mathcal{M}'_\psi = T_c^3\, (15 - \pi^2)/12\pi^2 . \tag{4.64}$$

Notice that for the sine pulse

$$\frac{\mathcal{M}'_\psi}{\mathcal{M}_\psi} = \frac{15-\pi^2}{15+2\pi^2} \approx 0.1477, \tag{4.65}$$

which is smaller than the ratio of these two parameters for rectangular pulses. For rectangular pulses, (4.60) and (4.61) imply that this ratio is exactly 0.5. As a result, the parameter $\mu_{k,i}(1)$ contributes less to the interference parameter $\sigma^2_{k,i}$ for the sine pulse than it does for the rectangular pulse. From (4.57) we see that for the sine pulse

$$\sigma^2_{k,i} = (12N^3\pi^2)^{-1}\{(15+2\pi^2)\mu_{k,i}(0) + (15-\pi^2)\mu_{k,i}(1)\} \tag{4.66}$$

$$= \beta_1(6N^3)^{-1}\{2\mu_{k,i}(0) + \beta_2\mu_{k,i}(1)\}, \tag{4.67}$$

where

$$\beta_1 = (4\pi^2)^{-1}(15+2\pi^2) \approx 0.8800 \tag{4.68}$$

and

$$\beta_2 = 2(15-\pi^2)(15+2\pi^2)^{-1} \approx 0.2954 . \tag{4.69}$$

For many sets of sequences of interest

$$|\mu_{k,i}(1)| << \mu_{k,i}(0) . \tag{4.70}$$

If the signature sequences satisfy (4.70) then

$$\sigma^2_{k,i} \approx \mu_{k,i}(0)/3N^3 \tag{4.71}$$

for rectangular pulses and

$$\sigma^2_{k,i} \approx \beta_1\mu_{k,i}(0)/3N^3 \tag{4.72}$$

for sine pulses. Thus, the first-order effect of using sine pulses instead of rectangular pulses is a decrease in the multiple-access interference by about 12%. For sets of signature sequences of interest the improvement is rarely as large as 12%. For AO/LSE phases of m-sequences of periods 31 and 63 (see [Pursley and Roefs (1979)] for a description of these sequences), the improvement is in the range 8%-11%. For phases of these sequences which are optimum with respect to $\sigma_{k,i}^2$ (see [Pursley, Garber, and Lehnert (1980)]), the improvement is in the range 7%-11%. There is only one possible explanation for the fact that the improvement is less than the 12% that is predicted by consideration of the leading terms in (4.62) and (4.67). The difference must be due to the second terms, and in order for the second terms to account for a reduction in the percentage improvement, some or all of the parameters $\mu_{k,i}(1)$ must be negative for these sets of sequences. Indeed, for all of the sequences mentioned above, $\mu_{k,i}(1) < 0$ for all $k \neq i$. Since $|\mu_{k,i}(1)|/\mu_{k,i}(0)$ is small, however, this is a second-order effect. The values of $\mu_{k,i}(0)$ and $\mu_{k,i}(1)$ for the AO/LSE phases of the three non-reciprocal m-sequences of period 31 are given in Table 1. The values of $\sigma_{k,i}^2$ for the rectangular pulse and for the sine pulse are given in Table 2. The value of $\sigma_{k,i}^2$ for the rectangular pulse can also be obtained from the data presented in [Pursley and Roefs (1979), Figure A.3], since the parameter $r_{k,i}$ tabulated there is related to $\sigma_{k,i}^2$ by (3.30).

It is certainly not true that $\mu_{k,i}(1)$ is always negative. For example, $\mu_{k,i}(1)$ is positive for several pairs of AO/LSE sequences of period 127. For certain pairs of these sequences the improvement actually exceeds 13%. Notice from (4.62) and (4.67) that the ratio of $\sigma^2_{k,i}$ for the sine pulse to $\sigma^2_{k,i}$ for the rectangular pulse must be less than β_1 whenever $\mu_{k,i}(1) > 0$, since $\beta_2 < 1$.

Table 1. Correlation Parameters AO/LSE m-Sequences of Period 31

k,i	$\mu_{k,i}(0)$	$\mu_{k,i}(1)$
1,2	967	-88
1,3	1015	-40
2,3	1015	-120

Table 2. Interference Parameters for AO/LSE m-Sequences of Period 31

k,i	$\sigma^2_{k,i}$ (sine pulse)	$\sigma^2_{k,i}$ (rectangular pulse)
1,2	0.009393	0.010328
1,3	0.009935	0.011133
2,3	0.009819	0.010686

V. QUATERNARY DIRECT-SEQUENCE SSMA COMMUNICATIONS

The signal model for Section II is binary direct-sequence spread-spectrum modulation in which the product of a baseband data signal

$$b_n(t) = \sum_{\ell=-\infty}^{\infty} b_\ell^{(n)} p_T(t-\ell T) \tag{5.1}$$

and a wideband signal

$$a_n(t) = \sum_{j=-\infty}^{\infty} a_j^{(n)} \psi(t-jT_c) \tag{5.2}$$

is the amplitude modulation for a carrier. The resultant spread-spectrum signal can be viewed as a binary amplitude-modulated signal or a binary phase-shift-keyed (PSK) signal. The pulse shapes are not required to be rectangular, however.

In the present section we consider signals which are given by

$$s_k(t) = s_k^Q(t) + s_k^I(t) \tag{5.3}$$

where $s_k^Q(t)$ is the quadrature component of the k-th signal and $s_k^I(t)$ is the in-phase component. The quadrature component is given by

$$s_k^Q(t) = A\, a_{2k-1}(t) b_{2k-1}(t) \sin(\omega_c t + \theta) \tag{5.4}$$

and the in-phase component is

$$s_k^I(t) = A\, a_{2k}(t-t_o) b_{2k}(t-t_o) \cos(\omega_c t + \theta) \tag{5.5}$$

where $a_{2k-1}(\cdot)$ and $a_{2k}(\cdot)$ are as in (5.2), $b_{2k-1}(\cdot)$ and $b_{2k}(\cdot)$ are as in (5.1), and t_o is the amount of offset between the in-phase and quadrature components of the signal $s_k(t)$. We assume that this offset is the same for all of the signals. For convenience we assume that

$t_o = \frac{1}{2}\nu T_c$ for some integer ν; that is, t_o is some multiple of one-half of the chip duration. This assumption is consistent with actual system implementations for which the most common choices are $t_o = 0$ or $t_o = \frac{1}{2} T_c$. The value $t_o = 0$ is often employed in standard quadriphase-shift-keying (QPSK), and $t_o = \frac{1}{2} T_c$ can be used for either minimum-shift-keying (MSK) or offset quadriphase-shift-keying (OQPSK). In case N is odd then $t_o = \frac{1}{2}T$ is a possible choice for the offset in either MSK or OQPSK, but in fact $t_o = \frac{1}{2}(2\nu + 1)T_c$ is suitable for MSK or OQPSK for any value of the integer ν. The results presented in this section are actually valid under more general restrictions on t_o; however, there appears to be no good reason for employing values of t_o other than multiples of $\frac{1}{2} T_c$.

The signal specified by (5.1)-(5.5) can be thought of as offset quadrature amplitude modulation or as a generalized form of QPSK modulation. The chip waveform $\psi(t)$ is of arbitrary shape, but as in Section II it is assumed to satisfy $\psi(t) = 0$ for $t \notin [0,T_c]$ and (by an appropriate normalization if necessary) to satisfy (2.2). As in Section II, the two primary examples are the rectangular pulse

$$\psi(t) = p_{T_c}(t) \qquad (5.6)$$

and the sine pulse

$$\psi(t) = \sqrt{2}\, \sin(\pi t/T_c)\, p_{T_c}(t) \; . \qquad (5.7)$$

If $\psi(t)$ is the rectangular pulse then the signal $s_k(t)$ is a QPSK signal for $t_o = 0$ (or more generally if $t_o = \nu T_c$ for some integer ν) or an OQPSK signal if $t_o = \frac{1}{2}(2\nu + 1)T_c$ for some integer ν. If $\psi(t)$ is the sine pulse and $t_o = \frac{1}{2}(2\nu + 1)T_c$ for some integer ν then $s_k(t)$ is an MSK signal.

The reader who is not familiar with MSK or OQPSK modulation may wish to consult one of the standard references on these modulation schemes such as the paper by Gronemeyer and McBride (1976); however, our treatment is self-contained and our approach is to obtain results for MSK and OQPSK spread-spectrum signaling as a special case of the more general signal structure of (5.1)-(5.4). We refer to signals of this general type as <u>generalized quadriphase</u> or <u>quaternary</u> direct-sequence spread-spectrum signals.

Our model for an asynchronous quaternary direct-sequence spread-spectrum multiple-access (QDS/SSMA) communications system is shown in Figure 2. In this model, the received signal consists of additive white Gaussian noise n(t) plus the sum of the K signals $s_k(t-\tau_k)$, $1 \le k \le K$, where τ_k is the relative time delay associated with the k-th signal. Thus, as illustrated in Figure 2, the received signal is given by

$$\begin{aligned} r(t) = n(t) &+ \sum_{k=1}^{K} A\, a_{2k-1}(t-\tau_k) b_{2k-1}(t-\tau_k) \sin(\omega_c t + \varphi_k) \\ &+ \sum_{k=1}^{K} A\, a_{2k}(t-t_o-\tau_k) b_{2k}(t-t_o-\tau_k) \cos(\omega_c t + \varphi_k) \qquad (5.8) \end{aligned}$$

where $\varphi_k = \theta_k - \omega_c \tau_k$ (mod 2π). As before, the noise spectral density is $\frac{1}{2}N_0$. This model is an adaptation of our model for asynchronous binary direct-sequence SSMA which is shown in Figure 1 of Section II.

We wish to consider the output of the receiver for the i-th signal $s_i(t)$. Since we are interested only in <u>relative</u> time delays and phase angles, we may assume that $\tau_i = 0$ and $\theta_i = 0$ as in Section II. For $\tau_i = \theta_i = 0$, the receiver for the i-th signal consists of a pair of correlation receivers as shown in Figure 3. The n-th correlation

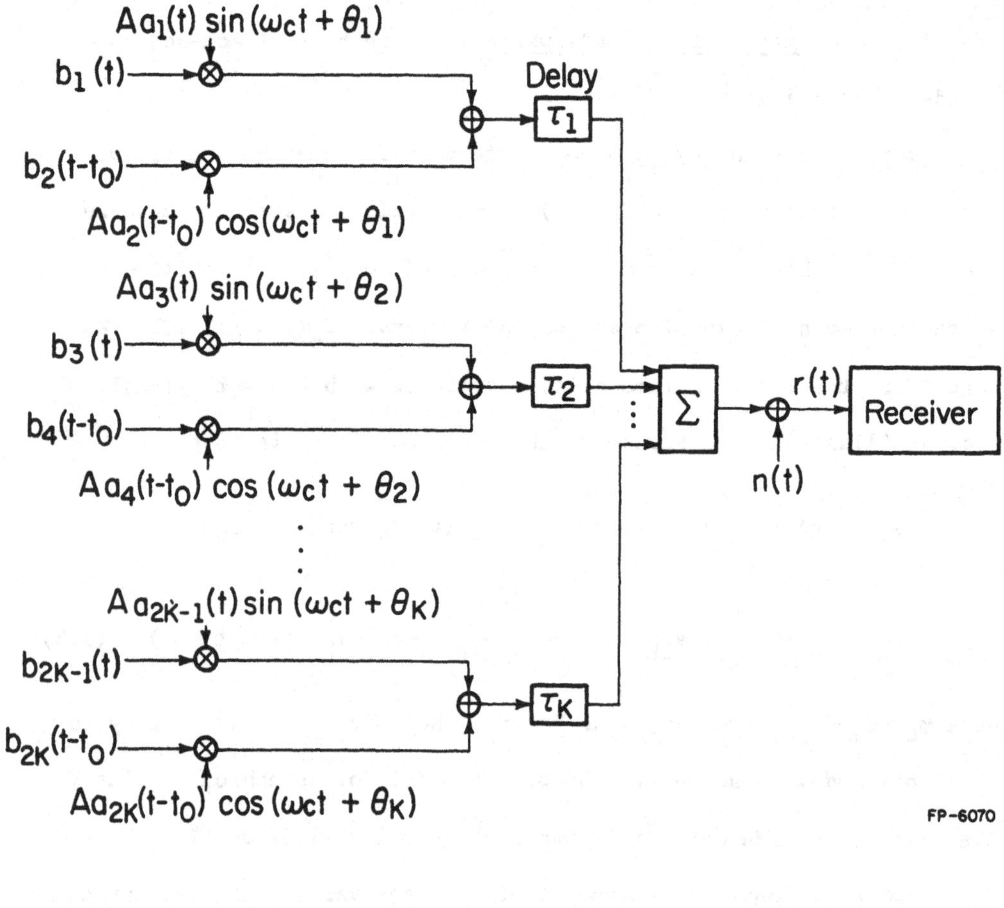

Figure 2. Quaternary Direct-Sequence SSMA Communications System Model.

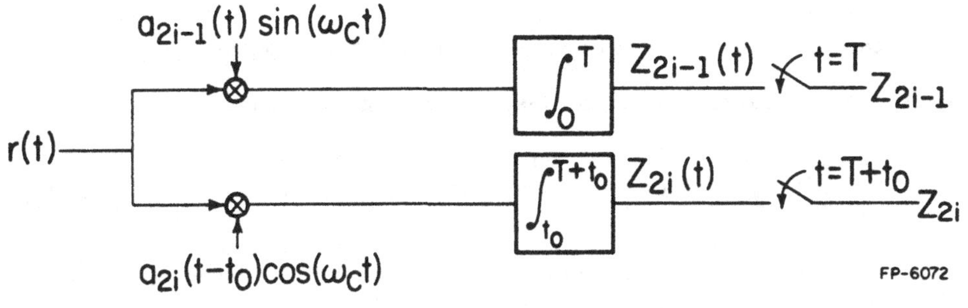

Figure 3. Receiver for Quaternary Direct-Sequence SSMA Communications.

receiver ($n=2i-1$ or $n=2i$) is matched to a spread-spectrum signal $c_n(t)$, which is often referred to as a phase-coded carrier. This signal is given by

$$c_n(t)=a_n(t)\sin(\omega_c t+\varphi_i) \tag{5.9}$$

for $n=2i-1$ (the quadrature carrier) and

$$c_n(t)=a_n(t-t_o)\cos(\omega_c t+\varphi_i) \tag{5.10}$$

for $n=2i$ (the in-phase carrier). Notice that if all double-frequency components are negligible then

$$\int_0^T c_{2i-1}(t)c_{2i}(t)dt=\int_{t_o}^{t_o+T} c_{2i-1}(t)c_{2i}(t)dt=0, \tag{5.11}$$

and hence there is no cross talk or self interference between the two components of the quaternary spread-spectrum signal. That is,

$$\int_0^T s_i^I(t)c_{2i-1}(t)dt=\int_{t_o}^{t_o+T} s_i^Q(t)c_{2i}(t)dt=0\ . \tag{5.12}$$

The effects of double-frequency components in an OQPSK spread-spectrum system are considered in [Garber and Pursley (1980)] where it is shown that they are negligible under certain conditions (such as if $\omega_c \gg T_c^{-1}$). In all that follows, the double-frequency components of the signals $r(t)c_n(t)$ are ignored. Thus, the quantities

$$Z_{2i-1}=\int_0^T r(t)c_{2i-1}(t)dt \tag{5.13}$$

and

$$Z_{2i}=\int_{t_o}^{t_o+T} r(t)c_{2i}(t)dt \tag{5.14}$$

can be written as (cf. (2.20))

$$Z_{2i-1} = \tfrac{1}{2}AT\{b_0^{(2i-1)} + \sum_{k\neq i} I_{2k-1,2i-1}(\underline{b}_{2k-1},\tau_k,\varphi_k) + \sum_{k\neq i} I_{2k,2i-1}(\underline{b}_{2k},\tau_k + t_o,\varphi_k)\} + \eta_{2i-1} \tag{5.15}$$

and

$$Z_{2i} = \tfrac{1}{2}AT\{b_0^{(2i)} + \sum_{k\neq i} I_{2k,2i}(\underline{b}_{2k},\tau_k,\varphi_k) + \sum_{k\neq i} I_{2k-1,2i}(\underline{b}_{2k-1},\tau_k - t_o,\varphi_k)\} + \eta_{2i} \tag{5.16}$$

where $\underline{b}_n$ is a vector of two consecutive data bits from $b_n(t)$. The quantities η_{2i-1} and η_{2i} are independent, zero-mean, Gaussian random variables, each of which has variance $\tfrac{1}{2}N_0T$. The quantities $I_{n,m}(\underline{b},\tau,\varphi)$ are defined as follows. Let $\underline{b}$, τ, and φ be such that

$$\underline{b} = (b,b'),\ b \in \{-1,+1\},\ b' \in \{-1,+1\},$$

$\tau \in [0,T]$, and $\varphi \in [0,2\pi]$. For $n = 2k-1$ and $m = 2i-1$, define

$$I_{n,m}(\underline{b},\tau,\varphi) = T^{-1}\{b \int_0^{\tau} a_n(t-\tau)a_m(t)dt + b' \int_{\tau}^{T} a_n(t-\tau)a_m(t)dt\}\cos\varphi. \tag{5.17}$$

Notice that the offset has no effect on <u>this</u> component of the interference. For $n = 2k$ and $m = 2i-1$,

$$I_{n,m}(\underline{b},\tau,\varphi) = T^{-1}\{b \int_0^{\tau+t_o} a_n(t-\tau-t_o)a_m(t)dt + b' \int_{\tau+t_o}^{T} a_n(t-\tau-t_o)a_m(t)dt\}\sin(-\varphi) \tag{5.18}$$

where $\tau + t_o$ is interpreted modulo $[0,T]$. If $n = 2k-1$ and $m = 2i$ then for $\tau \geq t_o$ we have

$$\begin{aligned} I_{n,m}(\underline{b},\tau,\varphi) &= T^{-1}\{b\int_{t_o}^{\tau} a_n(t-\tau)a_m(t-t_o)dt \\ &+ b'\int_{\tau}^{T+t_o} a_n(t-\tau)a_m(t-t_o)dt\}\sin\varphi \\ &= T^{-1}\{b\int_0^{\tau-t_o} a_n(t+t_o-\tau)a_m(t)dt \\ &+ b'\int_{\tau-t_o}^{T} a_n(t+t_o-\tau)a_m(t)dt\}\sin\varphi, \end{aligned} \tag{5.19}$$

but for $\tau < t_o$

$$\begin{aligned} I_{n,m}(\underline{b},\tau,\varphi) &= T^{-1}\{b\int_{t_o}^{T+\tau} a_n(t-\tau)a_m(t-t_o)dt \\ &+ b'\int_{T+\tau}^{T+t_o} a_n(t-\tau)a_m(t-t_o)dt\}\sin\varphi \\ &= T^{-1}\{b\int_0^{T+\tau-t_o} a_n(t+t_o-\tau)a_m(t)dt \\ &+ b'\int_{T+\tau-t_o}^{T} a_n(t+t_o-\tau)a_m(t)dt\}\sin\varphi. \end{aligned}$$

If $\tau - t_o$ is always interpreted modulo $[0,T]$ then (5.19) is valid for both $\tau \geq t_o$ and $\tau < t_o$. For future use, the difference $\tau - t_o$ in (5.19) is to be evaluated modulo $[0,T]$. Finally, for $n = 2k$ and $k = 2i$ we have

$$I_{n,m}(\underline{b},\tau,\varphi) = T^{-1}\{b \int_{t_o}^{t_o+\tau} a_n(t-t_o-\tau)a_m(t-t_o)dt$$

$$+ b' \int_{t_o+\tau}^{T+t_o} a_n(t-t_o-\tau)a_m(t-t_o)dt\}\cos\varphi$$

$$= T^{-1}\{b \int_0^{\tau} a_n(t-\tau)a_m(t)dt$$

$$+ b' \int_{\tau}^{T} a_n(t-\tau)a_m(t)dt\}\cos\varphi. \tag{5.20}$$

The multiple-access interference components given by (5.17)-(5.20) can be written in terms of the continuous-time partial crosscorrelation functions (defined in (2.17) and (2.18)) as follows. From (5.17) we see that for $n = 2k-1$ and $m = 2i-1$

$$I_{n,m}(\underline{b},\tau,\varphi) = T^{-1}\{b\ R_{n,m}(\tau) + b'\hat{R}_{n,m}(\tau)\}\cos\varphi. \tag{5.21}$$

Similarly, (5.18) implies that for $n = 2k$ and $m = 2i-1$

$$I_{n,m}(\underline{b},\tau,\varphi) = T^{-1}\{b\ R_{n,m}(\tau+t_o) + b'\hat{R}_{n,m}(\tau+t_o)\}\sin(-\varphi), \tag{5.22}$$

where by $\tau+t_o$ we mean $\tau+t_o$ modulo $[0,T]$ as mentioned above. In all that follows, it is assumed that the arguments of the continuous-time crosscorrelation functions $R_{n,m}$ and $\hat{R}_{n,m}$ are to be reduced modulo $[0,T]$. With this convention in mind, we see that (5.19) is equivalent to

$$I_{n,m}(\underline{b},\tau,\varphi) = T^{-1}\{b\ R_{n,m}(\tau-t_o) + b'\hat{R}_{n,m}(\tau-t_o)\}\sin\varphi. \tag{5.23}$$

for $n = 2k-1$ and $m = 2i$. Finally, (5.20) implies that for $n = 2k$ and $m = 2i$

$$I_{n,m}(\underline{b},\tau,\varphi) = T^{-1}\{b\ R_{n,m}(\tau) + b'\hat{R}_{n,m}(\tau)\}\cos\varphi. \tag{5.24}$$

In view of (5.15), (5.16), and (5.21)-(5.24), it is clear that our previous analysis of the binary system (Sections III and IV) applies also to the quaternary system. From this analysis we can obtain bounds on the interference (e.g., as in (4.49)), and we can evaluate the mean-squared interference which is necessary in order to determine the signal-to-noise ratio. Only the latter will be presented here.

We make the same assumptions regarding the distributions of the data symbols, phase angles, and time delays as in Section III. We assume that the K spread-spectrum signals are of equal power (this assumption was actually made in (5.4) and (5.5)), but the results are easily extended to unequal signal levels as in (3.31)-(3.35).

The above assumptions lead us to define (cf. equation (3.21))

$$\sigma^2_{n,m} = \mathrm{Var}\{I_{n,m}(\underline{b}_n, \tau_n, \varphi_n)\}. \tag{5.25}$$

From (5.21)-(5.24) we see that

$$\sigma^2_{n,m} = \tfrac{1}{2}T^{-2}\{\int_0^T T^{-1}[R^2_{n,m}(\tau) + \hat{R}^2_{n,m}(\tau)]d\tau\} \tag{5.26}$$

for each pair (m,n), which is analogous to (3.22). From (5.15) and (5.16) it follows that the signal-to-noise ratio can be expressed as

$$\mathrm{SNR}_m = \tfrac{1}{2}AT[\mathrm{Var}\{Z_m | b_0^{(m)} = +1\}]^{-\frac{1}{2}} \tag{5.27}$$

for $m = 2i-1$ or $m = 2i$, where

$$\mathrm{Var}\{Z_m | b_0^{(m)} = +1\} = \tfrac{1}{4}N_0T + \tfrac{1}{4}A^2T^2\,\Sigma_n^{(i)}\,\sigma^2_{n,m}. \tag{5.28}$$

The symbol $\Sigma_n^{(i)}$ is used to denote the summation over all n in the range

$1 \leq n \leq 2K$ except $n = 2i - 1$ and $n = 2i$, which are excluded. Thus

$$\Sigma_n^{(i)} \sigma_{n,m}^2 = \sum_{k \neq i} [\sigma_{2k-1,m}^2 + \sigma_{2k,m}^2] \tag{5.29}$$

for $m = 2i-1$ and $m = 2i$. From (5.27) and (5.28) we have

$$\mathrm{SNR}_m = \left\{ \frac{N_0}{A^2 T} + \Sigma_n^{(i)} \sigma_{n,m}^2 \right\}^{-\frac{1}{2}} . \tag{5.30}$$

Since the energy per data bit is

$$\mathcal{E}_b = \tfrac{1}{2} \int_0^T s_k^2(t)dt$$

$$= \int_0^T [s_k^I(t)]^2 dt = \tfrac{1}{2} A^2 T \tag{5.31}$$

then (5.30) can be written as

$$\mathrm{SNR}_m = \left\{ \frac{N_0}{2\mathcal{E}_b} + \Sigma_n^{(i)} \sigma_{n,m}^2 \right\}^{-\frac{1}{2}} , \tag{5.32}$$

which is valid for $m = 2i-1$ and $m = 2i$.

Equation (5.32) establishes the dependence of the signal-to-noise ratio on the variance $\sigma_{n,m}^2$ of the multiple-access interference. The analysis of Section IV, especially equations (4.50)-(4.57), shows that

$$\sigma_{n,m}^2 = \tfrac{1}{2} T^{-3} \{\hat{\mathcal{M}}_{n,m} + \mathcal{M}_{n,m}\}$$

$$= \tfrac{1}{2} T^{-3} \{2\mu_{n,m}(0)\mathcal{M}_\psi + 2\mu_{n,m}(1)\mathcal{M}_\psi'\}$$

$$= T^{-3} \{\mu_{n,m}(0)\mathcal{M}_\psi + \mu_{n,m}(1)\mathcal{M}_\psi'\}, \tag{5.33}$$

where $\mathcal{M}_\psi$ and $\mathcal{M}'_\psi$ are defined in (4.54) and (4.55), respectively. Thus the signal-to-noise ratio can be written as

$$\mathrm{SNR}_m = \left\{ \frac{N_0}{2\mathcal{E}_b} + T^{-3} \Sigma_n^{(i)} [\mu_{n,m}(0)\mathcal{M}_\psi + \mu_{n,m}(1)\mathcal{M}'_\psi] \right\}^{-\frac{1}{2}} . \tag{5.34}$$

In (5.34) the parameters $\mathcal{M}_\psi$ and $\mathcal{M}'_\psi$ depend only on the waveform $\psi(t)$, the parameters $\mu_{n,m}(0)$ and $\mu_{n,m}(1)$ depend only on the signature sequences and they can be computed from sequence autocorrelation functions as indicated in (4.59), and the remaining parameters $\mathcal{E}_b$, N_0, and T do not depend on either the waveform or the signature sequences. Numerical evaluation of (5.34) for QPSK or OQPSK requires a substitution for $\mathcal{M}_\psi$ and $\mathcal{M}'_\psi$ from (4.60) and (4.61) and an evaluation of the parameters $\mu_{n,m}(0)$ and $\mu_{n,m}(1)$ for the signature sequences of interest. For MSK, equations (4.63) and (4.64) give the appropriate values of $\mathcal{M}_\psi$ and $\mathcal{M}'_\psi$ to be employed in (5.34).

We close this section with some numerical results for QPSK and MSK spread-spectrum systems. These results are from [Pursley, Garber, and Lenhert (1980)]. As in that paper, we let $\{a^{(2k)} : 1 \le k \le K\}$ be a given set of binary sequences and let the sequences $\{a^{(2k-1)} : 1 \le k \le K\}$ be their reverses. That is, for each k

$$a_j^{(2k-1)} = a_{N-1-j}^{(2k)} \tag{5.35}$$

so that $a_0^{(2k-1)} = a_{N-1}^{(2k)}$, $a_1^{(2k-1)} = a_{N-2}^{(2k)}, \ldots, a_{N-1}^{(2k-1)} = a_0^{(2k)}$. For a discussion of the correlation properties of reverse sequences see [Sarwate and Pursley (1980)]. The key property for the evaluation of signal-to-noise ratio is that a sequence and its reverse have the same aperiodic autocorrelation function (this follows immediately from (4.5) with $k=i$). Therefore from (5.34) and (4.59) we see that if for each k, $a^{(2k-1)}$ is the reverse of $a^{(2k)}$, then $\sigma^2_{n,m}$ is a constant for all choices of $n \in \{2k-1,2k\}$ and $m \in \{2i-1,2i\}$. That is, if $\sigma^2(k,i) \triangleq \sigma^2_{2k,2i}$ then $\sigma^2_{n,m} = \sigma^2(k,i)$ for all such n and m. This implies that $SNR_{2i-1} = SNR_{2i}$ for $1 \leq i \leq K$. We then define

$$SNR = \min\{SNR_m : 1 \leq m \leq 2K\}, \tag{5.36}$$

which is the minimum signal-to-noise ratio for the 2K correlation receivers.

The values of the parameters $\mu_{n,m}(0)$ and $\mu_{n,m}(1)$ for the AO/LSE m-sequences of period $N=31$ are given in Table 1 of Section IV. The resulting values for $\sigma^2_{n,m}$ for the rectangular pulse and the sine pulse are given in Table 2. From this data we can evaluate the signal-to-noise ratio for quaternary direct-sequence SSMA communications for signature sequences which are AO/LSE m-sequences of period 31 and for an arbitrary chip waveform $\psi(t)$. We simply use Table 1, (5.34), (5.36), and the values of $\mathcal{M}_\psi$ and $\mathcal{M}'_\psi$ which are obtained from (4.9), (4.10), (4.54), and (4.55). For the special cases QPSK, OQPSK, or MSK it is simpler to use (5.32), (5.36), and the data given in Table 2. In Table 3 we present the values of SNR for QPSK (which are the same as for OQPSK) and for MSK for several values of the energy per bit to noise density ratio $\mathcal{E}_b/N_0$. The signature sequences are the AO/LSE m-sequences of period 31. Both SNR and $\mathcal{E}_b/N_0$ are given in decibels (dB) using the standard convention $(SNR)_{dB} = 20 \log_{10} SNR$ and $(\mathcal{E}_b/N_0)_{dB} = 10 \log_{10}(\mathcal{E}_b/N_0)$. Notice from (5.27)

$\mathcal{E}_b/N_0$(dB)	MSK	QPSK or OQPSK
10	10.5	10.3
12	11.5	11.2
14	12.3	12.0
16	12.8	12.5
18	13.2	12.9
20	13.5	13.1

Table 3. Signal-to-Noise Ratio in Decibels for AO/LSE m-Sequences of Period 31.

$\mathcal{E}_b/N_0$(dB)	MSK	QPSK or OQPSK
10	10.8	10.7
12	11.9	11.7
14	12.8	12.6
16	13.4	13.2
18	13.8	13.6
20	14.2	13.9

Table 4. Signal-to-Noise Ratio in Decibels for Optimal Phases of m-Sequences of Period 31.

that SNR is a ratio of signal amplitudes (e.g., signal voltage levels) whereas $\mathcal{E}_b/N_0$ is a power ratio. Of course $(SNR)_{dB}$ would not change even if we were to work with a signal-to-noise power ratio SNR', since we would then define $(SNR)_{dB} = 10 \log_{10} SNR'$.

Notice that MSK gives a larger signal-to-noise ratio than QPSK by 0.2 dB or more for each of the values of $\mathcal{E}_b/N_0$ given in Table 3. This is because of the smaller values of $\sigma^2_{n,m}$ for MSK than for QPSK (as shown in Table 2). From (5.32), (5.36), and Table 2 it is easy to see that in the limit as $\mathcal{E}_b/N_0 \to \infty$, the signal-to-noise ratio for MSK exceeds that for QPSK by $10 \log_{10}(1.1045) \approx 0.43$ dB. For the specific set of signature sequences employed, this is the largest possible difference between $(SNR)_{dB}$ for MSK and $(SNR)_{dB}$ for QPSK.

It is clear from (5.34) that the relative values of signal-to-noise ratio for MSK and QPSK depend on the signature sequences, and therefore we can increase the signal-to-noise ratio by careful selection of the signature sequences. For a given set of sequences, the choice of the phase or starting point for the signature sequence will influence the parameters $\mu_{n,m}(0)$ and $\mu_{n,m}(1)$. In general the optimal phase for MSK may be different than for QPSK because of the different weightings given to $\mu_{n,m}(0)$ and $\mu_{n,m}(1)$ by the MSK and QPSK pulse shapes (see (5.34) and (4.60)-(4.69)). In Table 4 we show the values of SNR that result when optimal phases of the $K = 3$ m-sequences of period 31 are employed [Pursley, Garber, and Lehnert (1980)]. We see from Table 4 that for the optimal phases of these m-sequences the improvement in SNR for MSK over QPSK is in the range 0.1 dB - 0.3 dB for values of $\mathcal{E}_b/N_0$ between 10 dB and 20 dB. In the limit as $\mathcal{E}_b/N_0 \to \infty$, the signal-to-noise ratio for MSK is larger than for QPSK by about 0.35 dB. Notice that the optimal phases give substantially improved signal-to-noise ratios over the AO/LSE phases. The improvement at $\mathcal{E}_b/N_0 = 12$ dB is 0.4 dB for MSK and 0.5 dB for QPSK. For very large values of $\mathcal{E}_b/N_0$ the improvement is approximately 0.81 dB for MSK and 0.89 dB for QPSK.

ACKNOWLEDGEMENTS

First and foremost I would like to express my appreciation to Professor Giuseppe Longo for the opportunity to present a series of lectures on spread-spectrum multiple-access communications at the 1979 CISM summer school and for his invitation to prepare my lecture notes for publication in this volume. I also wish to thank F. D. Garber for many valuable suggestions on various drafts of these notes. Most of the results presented here were obtained in research which was supported by the U. S. National Science Foundation (ENG 78-06630), the U. S. Army Research Office (DAAG 27-78-G-0114), and the U. S. Joint Services Electronics Program (N00014-79-C-0424).

REFERENCES AND SELECTED BIBLIOGRAPHY

D. R. Anderson and P. A. Wintz (1969), "Analysis of a spread-spectrum multiple-access system with a hard limiter," IEEE Transactions on Communication Technology, vol. COM-17, pp. 285-290.

H. Blasbalg (1965), "A comparison of pseudo-noise and conventional modulation for multiple-access satellite communications," IBM Journal, vol. 9, pp. 241-255.

D. E. Borth and M. B. Pursley (1979), "Analysis of direct-sequence spread-spectrum multiple-access communication over Rician fading channels," IEEE Transactions on Communications, vol. COM-27, pp. 1566-1577.

D. E. Borth, M. B. Pursley, D. V. Sarwate, and W. E. Stark (1979), "Bounds on error probability for direct-sequence spread-spectrum multiple-access communications," 1979 Midcon Proceedings, Paper No. 15/1, pp. 1-14.

G. R. Cooper and R. W. Nettleton (1978), "A spread-spectrum technique for high-capacity mobile communications," IEEE Transactions on Vehicular Technology, vol. VT-27, pp. 264-275.

R. C. Dixon, Editor (1976), Spread Spectrum Techniques, IEEE Press, New York.

F. D. Garber and M. B. Pursley (1980), "Performance of offset quadriphase spread-spectrum multiple-access communications," submitted to IEEE Transactions on Communications.

S. A. Gronemeyer and A. L. McBride (1976), "MSK and offset QPSK modulation," IEEE Transactions on Communications, vol. COM-24, pp. 809-820.

IEEE (1977), IEEE Transactions on Communications: Special Issue on Spread Spectrum Communications, vol. COM-25, no. 8, 1977.

J. L. Massey and J. J. Uhran, Jr. (1969), "Final report for multipath study," (Contract No. NAS5-10786), University of Notre Dame, Notre Dame, Indiana.

J. L. Massey and J. J. Uhran, Jr. (1975), "Sub-baud coding," Proceedings of the Thirteenth Annual Allerton Conference on Circuit and System Theory, pp. 539-547.

M. B. Pursley (1974), "Evaluating performance of codes for spread spectrum multiple access communications," Proceedings of the Twelfth Annual Allerton Conference on Circuit and System Theory, pp. 765-774.

M. B. Pursley (1977), "Performance evaluation for phase-coded spread-spectrum multiple-access communication -- Part I: System analysis," IEEE Transactions on Communications, vol. COM-25, pp. 795-799.

M. B. Pursley (1979), "On the mean-square partial correlation of periodic sequences," Proceedings of the 1979 Conference on Information Sciences and Systems, Johns Hopkins Univ., pp. 377-379.

M. B. Pursley and F. D. Garber (1978), "Quadriphase spread-spectrum multiple-access communications," 1978 IEEE International Conference on Communications, Conference Record, vol. 1, pp. 7.3.1-7.3.5.

M. B. Pursley, F. D. Garber, and J. S. Lehnert (1980), "Analysis of generalized quadriphase spread-spectrum communications," 1980 IEEE International Conference on Communications, Conference Record, to appear.

M. B. Pursley and H. F. A. Roefs (1979), "Numerical evaluation of correlation parameters for optimal phases of binary shift-register sequences," IEEE Transactions on Communications, vol. COM-27, pp. 1597-1604.

M. B. Pursley and D. V. Sarwate (1976), "Bounds on aperiodic cross-correlation for binary sequences," Electronics Letters, vol. 12, pp. 304-305.

M. B. Pursley and D. V. Sarwate (1977a), "Evaluation of correlation parameters for periodic sequences," IEEE Transactions on Information Theory, vol. IT-23, pp. 508-513.

M. B. Pursley and D. V. Sarwate (1977b), "Performance evaluation for phase-coded spread-spectrum multiple-access communication -- Part II: Code sequence analysis," IEEE Transactions on Communications, vol. COM-25, pp. 800-803.

M. B. Pursley, D. V. Sarwate, and W. E. Stark (1980), "On the average probability of error for direct-sequence spread-spectrum systems," Proceedings of the Fourteenth Annual Conference on Information Sciences and Systems, Princeton, NJ, to appear.

H. F. A. Roefs and M. B. Pursley (1977), "Correlation parameters of random binary sequences," Electronics Letters, vol. 13, pp. 488-489.

D. V. Sarwate and M. B. Pursley (1980), "Crosscorrelation properties of pseudorandom and related sequences," Proceedings of the IEEE, vol. 68, pp. 593-619.

J. W. Schwartz, J. M. Aein, and J. Kaiser, "Modulation techniques for multiple access to a hard-limiting satellite repeater," Proceedings of the IEEE, vol. 54, pp. 763-777.

K. Yao, "Error probability of asynchronous spread spectrum multiple access communications systems," IEEE Transactions on Communications, vol. COM-25, pp. 803-809, 1977.

SOME TOOLS FOR THE STUDY OF CHANNEL-SHARING ALGORITHMS.

Gabriel RUGET
Mathématiques
Université Paris-Sud (Centre d'Orsay
91405 ORSAY (France)

§ I. About the algorithms.

We will only discuss discrete-time algorithms , with messages of standard length, equal to the discretization step : this may imply some agglutinations in the process of message arrivals, but we will neglect them.

ALOHA : In case of conflict (several simultaneously emitted, hence lost, messages) the terminals involved wait during random independent times, with a fixed distribution law, before re-emitting their messages. The simplest to realize is the geometric law : each "blocked" terminal tosses heads or tails at each moment, with a fixed bias π , to know if it is going to re-transmit its message. Globally, the arrival process is considered to be a Poisson one (each message proceeds from a different terminal, which is an unrealistic hypothesis if one is concerned with a small number of terminals and a high output, and in any case is an unfavourable one because it reduces to the extreme mutual information between messages). One can follow the evolution of the number of blocked terminals against time (which is markovian if the time latency's distribution is geometrical). It can be ascertained that, as feeble as the message output may be, the system becomes blocked after more or less time : mathematically, the Markov chain $N(t)$ "number of blocked terminals" at time t is a transient one. We will come back in § 2 to the definition of the system's "time before block", and to its study (by analytical means or by fast simulations).

Stabilized ALOHA : If, at each t, each terminal could know the number $n(t)$ of blocked terminals, it could choose with an indexed bias $\pi(n)$ if it is going to re-emit. The $n(t)$ process of the system's "states" would then remain Markovian ; Fayolle and al. [F.G.L] prooved that certain indexations $(\pi(n) = C/n)$ lead to a stabilization of the channel, provided the output be $< 1/e$ (the $n(t)$ chain is positive-recurrent). In fact, no terminal knows $n(t)$, but all can evaluate it in various ways : Gelenbe and al. [B.G] gave realizable algorithms having the same performance as the unrealizable algorithm just described.

Near saturation (output near $1/e$) the mean time before successful transmission of a message may be a good evaluation test of the system. But the probability that a message wait more than a given bound, which is directly related to a probability of buffer overflow, maybe a better test ; we will indicate how to manage with in § 4 (See Cottrell-Fort-Malgouyres fore more details).

"Abramson's theorem" : Abramson noticed that, if the superposition process of "fresh" messages and re-emitted messages is supposed to be Poisson, the output of the channel cannot exceed 1/e. It is only accidentally that this bound coincides with the one obtained by the previous method : we will now show some other "random" procedures for which the bound is over-stepped.

Capetenakis-Gallager's algorithm and its stabilization.

The messages are grouped into consecutive batches, each batch being set-up by a process we can visualize as a walk on a tree. In the original algorithm, batch n° 1 is formed by the messages blocked at the

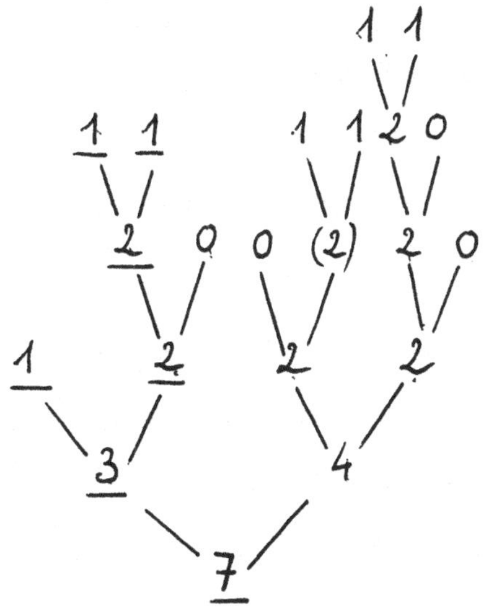

initial moment ; batch n° 2 by the messages generated during transmission of batch n° 1 and so on. Transmission of a batch (example of a size 7) : We describe it from a central view point, the next problem being to realize the procedure in a decentralized fashion. The batch is randomly divided into two parts (the right sub-batch will be transmitted after the first sub-batch on the left is completely transmitted) ; all the messages to the left are transmitted ; in case of success (emission of only one message) the right sub-batch is allowed to emit ; in case of failure (several messages) the left sub-batch is re-divided ; il the sub-batch allowed to re-emit happens to be empty, the right sub-batch is re-divided.

Realization : Let us define a level on the tree summits by deciding that the root is level 0, that the arrows to the right-hand side preserve the level, and that the arrows to the left-hand side increment it by one. Each terminal disposes of two counters ; a general counter $G(t)$, evolving in identical ways for all the terminals, who indicates the level of the node allowed to emit at time $t+1$; an individual counter $I(t)$, who indicates the level of the node attained by the terminal at time t (after emission and random subdivision). Here is the way to increment the counter at time $t+1$:

1°) if $I(t)=G(t)$, emission ; otherwise, silence.

2°) if success is attained, the setting becomes $G(t+1)=G(t)-1$ if a failure is observed, the setting becomes $G(t+1)=G(t)+1$; if silence is observed, and a failure was observed at the previous time, the setting becomes $G(t+1)=G(t)$.

3°) if a failure has just been ascertained, or if a silence is observed when a failure has been ascertained at the previous time, the setting becomes $I(t+1)=I(t)+1$ with a probability π , and $I(t+1)\neq I(t)$ with a probability $1-\pi$; otherwise, we have $I(t+1)=I(t)$.

Performances : The expectation τ_n of an n-size batch's transmission duration if given by the recurrent formula :

$$\tau_n = 1+\tau_n\ (1-\pi)^n + \sum_{k=1}^{n} \binom{n}{k} (\tau_k+\tau_{n-k})\ \pi^k\ (1-\pi)^{n-k}$$

$$\tau_o = \tau_1 = 1$$

For $\pi=1/2$, almost optimal value for the first τ_n (more than 1.5 % cannot be gained), the first τ_n/n values are :

1 2.25 2.33 2.41 2.46 2.49

The limit of τ_n/n can be calculated by randomizing n : let τ_μ be the expectation of the transmission duration of a poissonian size batch with an expectation of μ . Let us write :

$$\xi(x) = 1-e^{-x} - xe^{-x}$$

$$\tilde{\xi}(y) = \xi(y) - \frac{1}{2}\ e^{-y/2}\ \xi(y/2)$$

We have :

$$\tau_\mu = 1 + \sum_{k=0}^{\infty} 2^{k+1}\ \tilde{\xi}\ (\mu/2^k) \quad \text{and} \quad \tau_\mu/\mu \underset{\mu \to \infty}{\to} 2.68$$

On the other hand, μ/τ_μ goes through a maximum whose value is 0.463 when μ is near 1.3.

Capetenakis's original algorithm can be studied if we consider the T_i series of the resolution duration of successive batches ; if λ is the output of the Poisson message source, this series is a Markov chain, and

$$E\,(T_{i+1} - T_i / T_i) = \tau\,(\lambda T_i) - T_i$$

It follows that the chain is recurrent (stable system) if and only if $\lambda < 1/2.68$, which is a better bound that $1/\theta$.

A first improvement consists in solving batches of optimal (random) size : the number N of waiting messages is estimated ; each waiting message stores itself with a probability $\mu_0/\hat{N}$ in a batch which will be either activated or stay in waiting ; considering the batch's transmission duration and its size (observed a posteriori) , the estimation of N is re-actualized, ... It can be prooved that this procedure is stable for an input rate ≤ 0.463

From now to the end of this §, we will describe several improvements to the Capetenakis-Gallager algorithm, which seem to have been found first by Gallager but found independently by other workers in the field (russian and french ones, for instance J. Pellaumail, at INSA of Rennes).

From 0.463 to 0.487

The transmission of a batch stops when two consecutive successes have been observed (see underlined nodes in Fig. 1) : to state this differently, every sub-batch left to the right will be reconsidered only if its brother to the left has a size ≤ 1. Indeed, if ℓ is a Poisson variable whose expectation is E, the procedure dividing ℓ as d+g, when conditioned by $\ell > 1$, d is no more a Poisson variable ; but conditioned by $g > 1$, d is a Poisson variable whose expectation is E/2 ; it is then legitimate to send back the batch d on which no new information has been acquired to the reserve of blocked messages.

Evaluation : Let t_n be the expectation of the duration of a size n batch's partial transmission, and s_n the expectation of the number of transmitted messages. We have $s_o = 0$, $s_1 = 1$, $s_2 = 2$, $t_o = 1$, $t_1 = 2$, $t_2 = 4$, and the recurrence formulas (the summations are empty for n = 3)

$$s_n = \left(\frac{1}{2}\right)^n \left[2s_n + n\,(1 + 2s_{n-1}) + \sum_{k=2}^{n-2} \binom{n}{k} s_k\right]$$

$$t_n = \left(\frac{1}{2}\right)^n \left[2\,(1+t_n) + n\,(3+2t_{n-1}) + \sum_{k=2}^{n-2} \binom{n}{k} (1+t_k)\right]$$

n	3	4	5	6	7	8	9	10
s_n	2.5	2.57	2.52	2.50	2.50	2.50	2.50	2.51
t_n	5.83	6.48	6.67	6.84	7.03	7.22	7.39	7.54

If we randomize n by substituting a Poisson variable with an expectation μ , we find that the maximum of the ration s_μ/t_μ is 0.487 and is attained for $\mu = 1.27$ (this for subdivisions into batches of the same expect of size ; almost nothing is gained by introducing a bias into these subdivisions).. The evolution of the reserve of blocked message's size N_i at the beginning of each "group of transmissions" is markovian. If the source output is λ and if the treated batches are poissonian, with an expectation μ , we have :

$$E (N_{i+1} - N_i) = \lambda\, t_\mu - s_\mu$$

(for N_i large, otherwise it will be difficult to suppose that the transmitted batch is going to be poissonian !).

We can deduce recurrence if $\mu = 1.27$ and $\lambda < 0.487$.

A deterministic procedure : (see Berger's lecture in the present book). This last procedure tries to reduce the dispersal of the message's waiting times. It consists in an alternative to the method of batch subdivision by tossing heads or tails, but considered from the point of view of global output (stability, ...) it is identical to the algorithms quoted.

The terminals are ordered in a ring (virtual, but the order is supposed to be known of everyone). We suppose they are numerous, each contributing a small part to the global output. The ring is parametered by R modulo 1 in such a way as the global output of the terminals located on an arc be proportional to the measure of this arc (the arrival of fresh messages into the arc α is a Poisson process, with expectation $\lambda\alpha$).

A transmission algorithm covers the ring always in the same way, operating on a set of messages of optimized Poisson size, emptying provisionally an arc A_iA_{i+1} before moving on at time t_{i+1}, to an arc $A_{i+1}B_{i+1}$ to empty the contents of $A_{i+1}A_{i+2}$ at that time, ... Each terminal has in memory the history (A_i,t_i) of the groups of transmission during the preceding ring turn. The optimal B_j can therefore be calculated (for each element of arc following A_j, it is known for how much time the Poisson arrivals of known output must be integrated), and then, by listening to the transmissions (silence / success / failure), the instant t_{j+1} when a new batch must be reconstituted and the locus A_{j+1} where one must start again can be determined.

Let T_i be the time a message arrived at A_i has to wait just after the passage of the algorithm immediately preceding the passage taking place at time t_i (read it again !). To evaluate T_{i+1} from T_i in an approximative way (an exact model implies using a Markov chain on a function space) one can suppose that, conditionally to T_i, at time t_i, the messages that have just missed the algorithm on its preceding passage over A_{i+1} have been waiting for it for a time $T_i\ (1-A_iA_{i+1})$; since $A_iB_i = \frac{\mu}{\lambda T_i}$, one has $A_iA_{i+1} = a_i \frac{\mu}{\lambda T_i}$, where a_i is the ratio of the treated arc to the arc initially considered by the algorithm. Thus, $T_{i+1}-T_i = (t_{i+1}-t_i)\ \frac{\mu}{\lambda}\ a_i$.

Indeed we just consider this as an indication that will help us to determine the best way of introducing a bias into the procedure, for a fast simulation giving the probability of long waiting times (see § 4).

§ II. Destabilization of ALOHA ; introduction to large deviations.

Coming back to non-controlled ALOHA, let's look at

= E (N(t+1)-N(t) / N(t)=n) (We already computed it to prove transience). Typically, f(n) has the same sign as $(n-n_o)(n-n_c)$, and the important thing to remark is that the size of the process's jumps (the expectation and standard deviation of $N(t+1)-N(t)|N(t)=n$) is small compared to the variation n changing significantly this conditionned jumps'law $P_{n,.}$ (for instance n_c-n_o). So we don't consider this chain alone, but as being part of a fixed-drift family with homothetically smaller and smaller jumps, which can be studied asymptotically. To realize this family, time and space must be made continuous. Let us introduce a parameter $0 < \varepsilon \leq 1$, the process $X^\varepsilon(t)$ jumps at times in $\varepsilon\mathbb{N}$, with the transition probabilities

$$P^\varepsilon_{x,x+\varepsilon j} = P_{x,x+j} \qquad \text{(interpolate between the } P_{n,n+j}\text{)}$$

We will now give results for $\varepsilon \to 0$; according to the initial remark, they give good indications on the case $\varepsilon = 1$.

Law of large numbers : T fixed, $X^\varepsilon(0) = x_o$ fixed ; the trajectories $X^\varepsilon(t)$, $0 \leq t \leq T$, converge uniformly, with a probability converging to 1, to the trajectory stemming from x_o of the differential equation

$$\frac{dX}{dt} = f(X) = \sum_j j\, P_{x,x+j}$$

For instance, for every neighbourhood V of n_o, there exists an

S so that, for every $T > S$, the trajectories $X^{\varepsilon}(t)$, $S \leq t \leq T$, remain inside V with a probability converging to 1 when ε converges to 0 (cf. Kashminski).

Central limit theorem : T fixed, $X^{\varepsilon}(0) = n_o$ for simplification, the process $1/\sqrt{\varepsilon}\,(X^{\varepsilon}(t)-n_o)$, $0 \leq t \leq T$, converges, when $\varepsilon \to 0$, to the Ornstein-Uhlenbeck process

$$d\xi = -b\xi dt + \sigma dw$$

where $-b = f'(0)$, σ = standard deviation of the P_{n_o},.

Cramer's theorem (Chernoff) : δ fixed small, θ fixed small (but independent of ε), let π^{ε} be the probability of the set of trajectories (stemming from x_o fixed) so that

$$\forall t \, , \, 0 \leq t \leq \theta \, , \, |X^{\varepsilon}(t) - x_o - vt| \leq \delta$$

We have $- \varepsilon \log \pi^{\varepsilon} \underset{\varepsilon \to 0}{\to} \theta\, h_{x_o}(v)$

where h is the dual function of $L_{x_o}(\lambda)$ = Log (Laplace transform of $P^1_{x_o}$ at λ) :

$$- h(v) = \inf_{\lambda} L(\lambda) - \lambda v$$

NB : The gross character of the evaluation becomes evident if one considers the fact that δ is not included in the second term.

Action integrals ([A-R], [V]) : δ fixed small, T and $\varphi(t)$: $[0,T] \to \mathbb{R}$ fixed, let π^{ε} be the probability of the set of trajectories stemming from $\varphi(0)$ so that

$$\forall t\ ,\quad 0 \leq t \leq T,\quad |X^{\varepsilon}(t) - \varphi(t)| \leq \delta$$

We have $-\varepsilon \log \pi^{\varepsilon} \underset{\varepsilon \to 0}{\to} \int_{o}^{T} dt\, h_{\varphi(t)}\,(\varphi'(t)) = I(\varphi)$

Corollary : Let $n_o < y < x_o < z \leq n_c$, and let π^{ε} the probability of the set of trajectories stemming from x_o who exit from $[y,z]$ through z. We have :

$$-\varepsilon \log \pi^{\varepsilon} \underset{\varepsilon \to 0}{\to} \inf_{\substack{T,\varphi \\ \varphi(0)=x_o \\ \varphi(T)=z^o}} I\,(\varphi)$$

It can be easily seen that the infimum is reached when φ satisfies of the differential equation $\frac{d\varphi}{dt} = \chi(\varphi)$ (see figure) and that $\lambda(\varphi)$ is a root of $L(\lambda(\varphi)) = 0$

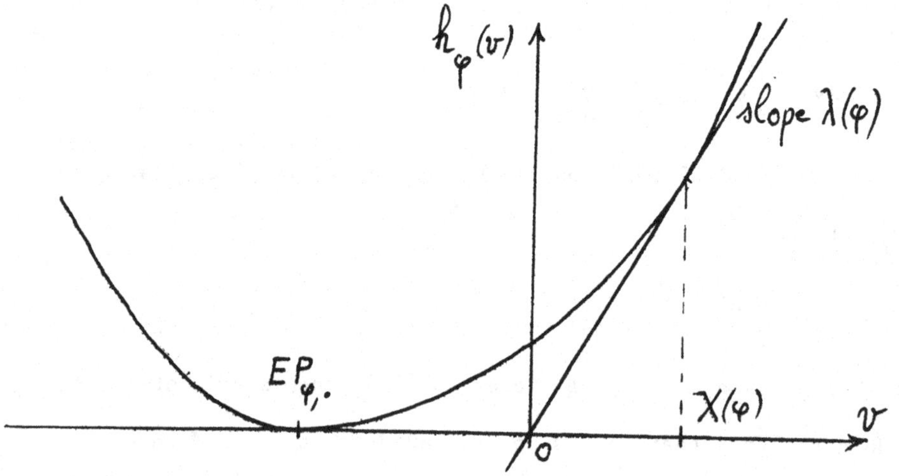

And therefore we have : $-\varepsilon \log \pi^{\varepsilon} \underset{\varepsilon \to 0}{\to} \int_{o}^{z} \lambda(x)\, dx$.

Interpretation : It is important to recall the proof of Cramer's theorem : the over-bounding of π^{ε} is well-known (Chernoff's inequality). The minoration can be obtained by changing the P^{ε} probabilities into $\tilde{P}^{\varepsilon}$ obtained from the laws

$$\tilde{P}_{x,y} = \frac{1}{\text{Laplace}\ (\lambda)}\ \exp \lambda\ (y-x)\ P_{x,y}$$

where λ maximises $\lambda v - L(\lambda)$. If so, the law of large numbers states that $\tilde{\pi}^{\varepsilon}$ is almost equal to 1 and one can easily bound, over the set of trajectories which interests us, the Radon-Nikodym derivative $dP^{\varepsilon} / d\tilde{P}^{\varepsilon}$.

Going back to the corollary, we can deduce from this that the "typical" trajectories of the $\tilde{P}^{\varepsilon}$ process constitute a subset of the set of trajectories exiting through z, whose probability is almost equal to π^{ε}, for $\tilde{P}_{x,y} = \exp\ (\lambda_{(\varphi)}\ (y-x))\ P_{x,y}$.

We are thus through with philosophy, and we can go on to methods of simulation and analytical calculus of the expectation τ of the first time of passage through n_c.

Fast simulation : We can consider $X(0) = n_o$. Every single trajectory stemming from n_o is decomposed into parts going back to n_o and into a last part joining n_o to n_c without going back through n_o. Evaluations are :

1°) by simulation with the $\tilde{P}_{i,j}$ probabilities, the P-probability p for a trajectory stemming from n_o to pass through n_c before passing again through n_o : let $X^i(t)$ be the simulated parts, stopped at their natural terms, either n_o or n_c. One can easily compute

$$\ell_i = \log \frac{dP}{d\tilde{P}} (X^i) = \sum_t \lambda (X^i(t)) (X^i(t+1)-X^i(t))$$

One estimator for p is

$$\hat{p} = \frac{\sum_{\text{Final } X^i=n_c} \exp \ell_i}{\sum_X \exp \ell_i}$$

2°) the mean time μ of return to n_o, by simulation with the $P_{i,j}$ probabilities, or by any other method (computations on a truncated Markov chain). Then take $\hat{\tau} = \hat{\mu}/\hat{p}$.

Approximations by diffusions : If 1 is considered to be very near 0, one can proceed as follows :

1°) evaluation of the P-probability, divided by x-y, of exiting from $[y,n_c]$ through n_c while starting from x : P is approximated by the homogeneous diffusion having at x the same drift and the same local variance, for which the result is analytically known (do $n_c = \infty$), that is to say $2b (x-n_o) / \tilde{\sigma} (x)$, where $\tilde{\sigma} (u)$ denotes the variance of $\tilde{P}_{u,\cdot}$.

2°) evaluation of $\log \frac{dP}{d\tilde{P}}$ following the typical $\tilde{P}$ trajectories: let $\phi (u) = \int_{n_o}^{u} \lambda (u) du$, and L the "typical" Radon-Nikodyon log-derivative. We have

$$\phi (X(t+1)) - \phi (X(t)) \simeq \lambda (X(t)) \Delta X + \frac{1}{2} \lambda' (X(t) (\Delta X)^2 ,$$

whence,
$$\phi (n_c) - \phi (x) \simeq - L + \frac{1}{2} \Sigma \lambda' (X(t)) \tilde{\sigma} (X(t))$$
$$\simeq - L + \frac{1}{2} \int \lambda' (\varphi(t)) \tilde{\sigma} (\varphi(t)) dt$$

where φ is the solution of $\varphi(0)=0$, $d\varphi/dt = \chi(\varphi)$;

$$\text{So,} \quad L = \phi(n_c) - \phi(x) + \int_x^{n_c} \frac{d\lambda}{2} \left(\frac{\frac{\partial^2 \hat{\mu}}{\partial \lambda^2}(\lambda,x) - [\frac{\partial \hat{\mu}}{\partial \lambda}(\lambda,x)]^2}{\frac{\partial}{\partial \lambda} \hat{\mu}(\lambda,x)} \right)_{\lambda=\lambda(x)}$$

3°) At this stage, we know $q = \frac{1}{x-y} P$ (starting from x, pass through n_c before passing through y)

$$q = \frac{2 \exp \phi(x) \chi(n_c)}{\sqrt{\tilde{\sigma}(x)} \exp \phi(n_c)} \frac{b(x-n_o)}{\chi(n_c)} \exp \int_x^{n_c} \frac{dx}{2} \frac{\partial \hat{\mu}}{\partial x} \left[1 - \frac{\frac{\partial^2 \hat{\mu}}{\partial \lambda^2}}{(\frac{\partial \hat{\mu}}{\partial x})^2}\right]_{\lambda=\lambda(x)}$$

$$= \frac{2 \exp \phi(x) \chi(n_c)}{\sqrt{\tilde{\sigma}(x)} \exp \phi(n_c)} \exp \int_{n_o}^{n_c} \frac{dx}{2} \left[\hat{\mu}_x + \hat{\mu}_\lambda \frac{\hat{\mu}_{\lambda\lambda}}{(\hat{\mu}_\lambda)^2} - \frac{2 \hat{\mu}_{\lambda x}}{\hat{\mu}_\lambda}\right]_{\lambda=\lambda(x)}$$

$$\text{because} \quad \frac{d\chi}{\chi} = dx \left[\frac{\frac{\partial^2 \hat{\mu}}{\partial \lambda \, \partial x}}{\frac{\partial \hat{\mu}}{\partial \lambda}} - \frac{\frac{\partial \hat{\mu}}{\partial x} \cdot \frac{\partial^2 \hat{\mu}}{\partial \lambda^2}}{(\frac{\partial \hat{\mu}}{\partial \lambda})^2} \right]$$

Neglecting the time of passage from x to y and from x to n_c, we can then assimilate the process to a diffusion (indeed an Orstein-Uhlenbeck process) to the left of x, with, as boundary condition at x, a mixture of death and reflection : the expectation $\tau(u)$ of the extinction time with start at u (which takes the place of the passage time at n_c) verifies

$$\begin{cases} \frac{\sigma}{2} \tau'' - b\,(u-n_o)\, \tau' = -1 & \sigma = \sigma(n_o) \\ \\ q\tau(x) + (1-q)\, \tau'(x) = 0 \end{cases}$$

From the first relation we can deduce :

$$\tau' \exp - \frac{b}{\sigma}(u-n_o)^2 = -\frac{2}{\sigma}\int_{-\infty}^{u} \exp - \frac{b}{\sigma}(v-n_o)^2 \, dV$$

If x is chosen large compared to $\sqrt{\frac{\sigma}{b}}$, but small enough so as vary little between 0 and x, we have

$$\frac{\tau'(x)}{\tau(x)} \simeq \frac{-2}{\tau(0)} \exp \frac{b(x-n_o)^2}{\sigma} \sqrt{\frac{\pi}{b\sigma}}$$

Whence $\tau(0) \simeq \dfrac{2 \exp \frac{b}{\sigma}(x-n_o)^2}{q} \sqrt{\dfrac{\pi}{b\sigma}}$

$$= \sqrt{\frac{\pi\sigma}{b}} \frac{\exp \phi(n_c)}{\hat{\mu}_\lambda(n_c,\lambda(n_c))} \exp \int_{n_o}^{n_c} dx\Big[\frac{\hat{\mu}_{\lambda x}}{\hat{\mu}_\lambda} \frac{\hat{\mu}_x}{2}\Big(1+\frac{\hat{\mu}_{\lambda\lambda}}{(\hat{\mu}_\lambda)^2}\Big)\Big]_{\lambda=\lambda(x)}$$

Remark 1 : If the process is a true diffusion, let's say $d\xi^\varepsilon = -b(\xi)\,dt + \sqrt{\varepsilon}\,dw$, where, in the vicinity of 0, we have $b(\xi) \sim b\xi$, the same method gives the expectation of the time of exit from an interval $[A,B]$.

For example, in the symmetrical case

$$\tau \sim \frac{1}{2b(B)} \sqrt{\frac{\varepsilon\pi}{b}} \exp \frac{\phi(B)}{\varepsilon}$$

a result prooved by Ludwig with a completely different method (the two methods pass -as difficultly or as easily- to the multidimensional case).

Remark 2 : If the continuous approximation is judged too gross (in the first and third step, more specifically) one can always replace it by computation on an ad hoc Markov chain with a small number of states.

Remark 3 : The justification of the approximations of the second step implies knowing how to over-bound the $\tilde{P}$-probability of the trajectories at a certain "distance" from the $\tilde{P}$-typical trajectories ; we will give an idea of the available tools in § 3.

The interested reader if referred to [C.F.M] for results of an experimentation along these lines.

APPENDIX : Justification of the choice of our fast simulation :

p is obtained as the mean of an independent N-sample of the image law of $\tilde{P}$ by $dP/d\tilde{P}$. Let us allow ourselves all the $\tilde{P}$ associated to exponential transformations

$$\tilde{P}_{x,y} = \frac{1}{\text{Laplace}_x\,(l_x)} \exp l_x\,(y-x)\; P_{x,y}$$

To be more precise, we shall write $\tilde{P} = P^l$, where $l : \{x\} \to \mathbb{R}$.
Let us search for l minimizing the variance of the estimator $\hat{p}$, that is to say

$$\int_{\text{trajectories exiting through } n_c} \left(\frac{dP}{dP^l} \right)^2 dP^l \quad .$$

Asymptotically, the preceding integral can be replaced by the maximum value of the integrand, or

$$\max_{\varphi : [0,T] \to [n_o, n_c]} \exp \int -2\,(l_\varphi d\varphi - L_\varphi\,(l_\varphi)\,dt) - h_\varphi^{l_\varphi}\,(\dot{\varphi})\,dt$$

where $h_\varphi^{l_\varphi}$ is the Cramer transform of the measure $P_{\varphi(t),\cdot}^{l_\varphi(t)}$ which is to say the dual of Log $\hat{P}_\varphi\,(p+l_\varphi)$ - Log $\hat{P}_\varphi\,(l_\varphi)$, or

$$h_\varphi^{l_\varphi}(v) = h_\varphi(v) + L_\varphi(l_\varphi) - v l_\varphi .$$

The log-variance of the estimator associated to l "is" then

$$\max_\varphi \int_{n_o}^{n_c} \left[\frac{-h_\varphi(\dot\varphi) + L_\varphi(l_\varphi)}{\dot\varphi} - l_\varphi \right] d\varphi$$

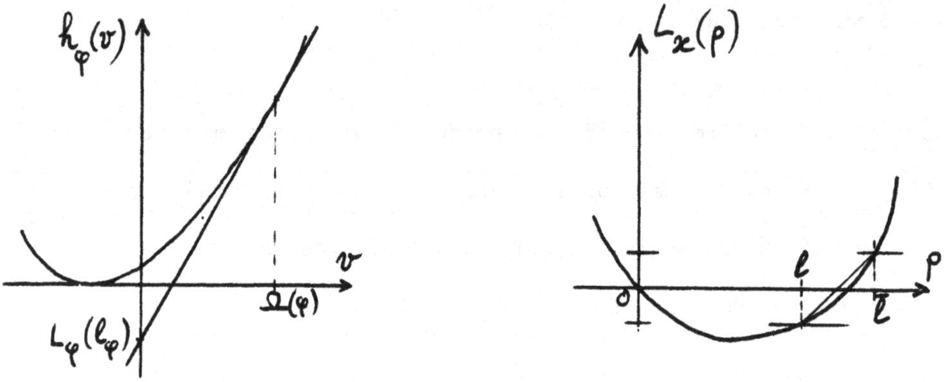

The maximizing φ is a solution of the differential equation $\dot\varphi = \Omega(\varphi)$ (see figure), the corresponding value of $\frac{-h_\varphi(\dot\varphi) + L_\varphi(l_\varphi)}{\dot\varphi}$ being $-\bar{l}_\varphi$, the larger of the two roots of the equation

$$L_\varphi(\bar{l}_\varphi) = - L_\varphi(l_\varphi)$$

(if the equation has no roots, $\max_\varphi = +\infty$ and l_φ is particularly ill-chosen).

Finally, we must minimize with respect to l_φ : this is done point by point : on x, we minimize $-\bar{l}_x - l_x$, and according to the convexity of L_x, this is obtained for l_x so that $L_x(l_x) = 0$.

§ 3. Large deviations (continued).

We first give the method for building large deviations theorems in dependent situations, for empirical laws. For a more profound study of these questions, we adress you to [D-V], [M] and [G]. For a few suggestions on applications, to [R]. We indicate the formulas concerning the Markov chains over Z^k with periodical transition probabilities ($P_{i,j} = P_{\underline{i},\, j-i}$, where $\underline{i}$ indicates the class of i modulo a net), which as far as we know are not to be found in the lïtterature : this is the beginning of an answer to a question asked in [BLP].

Large deviations for Markov chains.

In § 2, we hinted at the generic outline of the proof of a large deviation result : starting with a local study, with "frozen" laws, we went on from local gross results (equivalent of probabilities 'logarithms) to global gross results ; or we used the probability changes suggested by the local study for a global precise study. Here, we will enlarge the range of local available results : let us consider a process over R^k (in discrete time, to make things simpler), with small increments whose process is 1-markovian (here too, we could generalize).

$$P\,(X_{n+1} = y \,/\, X_n = x,\ X_{n-1} = w, \dots) = P\,(x;\ \frac{x-w}{\varepsilon}\ ;\ \frac{y-x}{\varepsilon})$$

and let us study the probability of an excentric mean increment over a time $1/\varepsilon$. Our aims being purely pedagogic, let us suppose the set of authorized incrementations finite : we therefore start from a finite set I, furnished with an application X to R^k and (irreducible) transition probabilities $P_{i,j}$. For every $\gamma \in (\mathbb{R}^k)^*$ we can build new probabilities $\tilde{P}_{i,j}$

$$\tilde{P}_{i,j} = P_{i,j} \exp(-\alpha + \beta_j - \beta_i + \gamma X_j) \quad .$$

The normalisation conditions $\Sigma \tilde{P}_{ij} = 1$ become

$$\sum_j P_{ij} \exp \gamma X_j \exp \beta_j = \exp \alpha \exp \beta_i$$

and this determinates the α and the β_i : $\exp \alpha$ is the eigenvalue of $(P_{ij} \exp \gamma X_j)$ associated to the positive eigenvector $(\exp \beta_i)$. The reason for this probability change is that, if $i_o, \ldots, i_N$ is a trajectory of the chain,

$$P/\tilde{P} \text{ (trajectory } i_o, \ldots, i_N) = \exp (N\alpha - \beta_{i_N} + \beta_{i_o} - \gamma \sum_{\ell=1}^{N} X_i)$$

so that, for every trajectory such that $\frac{1}{N} \sum_{\ell=1}^{N} X_{i_\ell} \simeq v$

$$\frac{1}{N} \log \frac{P}{\tilde{P}} \text{ (length N trajectory, mean speed v)} \simeq \alpha(\gamma) - \gamma v$$

When γ varies, $P(N,v)$ does not change, $\alpha(\gamma)-\gamma v$ passes through a minimum when $\tilde{P}(N,v)$ passes through a maximum. We cannot hope for this last one to exceed 1, and of course if we consider only the v belonging to the interior of the convex envelope of $\{X_i\}$. Reciprocally (if all $P_{ij} > 0$), using the ergodic theorem for $\tilde{P}_{ij}$, and by showing that, if γ varies, all the desirable v can be attained (Brouwer's theorem used in a recurrent way on the skeleton of the convex hull of $\{X_i\}$), we can see that :

$$\frac{1}{N} \log P(N,v) \gtrsim \sup_{\gamma} \alpha(\gamma) - \gamma v$$

In the finite case, by taking an application X so that the X_i be linearly independent, we can move on directly to a statement on the empirical laws of the i. Let ν be a probability on I :

we can write the limit $h_p(\nu)$ of $-1/N \log P(N,\nu)$, where $P(n,\nu)$ is the probability that the empirical law of a trajectory of length N be near ν, as

$$\sup_{\gamma} \int \log \frac{u_\gamma}{Pu_\gamma} \, d\nu \quad ,$$

where $u_\gamma(j) = \exp(\beta_j + \gamma_j - \alpha)$

whence

$$h_p(\nu) \leq \sup_{u>0} \int \log \frac{u}{Pu} \, d\nu$$

If w is a positive function over $I \times I$, by writing $Pw(i) = \sum_j P_{ij} w(i,j)$, and if λ is any every law on $I \times I$ having its two margins equal to ν, we have

$$\forall\lambda \quad , \quad \sup_{u>0} \int \log \frac{u}{Pu} \, d\nu \leq \sup_{w>0} \int \log \frac{w(i,j)}{Pw(i)} \, d\lambda$$

(take $w(i,j) = u(j)$).

With a minimax theorem

$$\sup_{u>0} \int \log \frac{u}{Pu} \, d\nu \leq \sup_{w} \inf \int \log \frac{w}{Pw} \, d\lambda$$

Finally, let us show that the term to the right is over-bounded by the infimum of $-1/N \log P(N,\nu)$ and this will give us two expressions of $h_p(\nu)$, which can be generalized to the case I compact and beyond (cf. [D-V]) and what is more, the interpretation of $h_p(\nu)$ as true limit. We begin by noticing that

$$E\left(\exp \sum_{\ell=0}^{N-1} \log \frac{w(i_\ell, i_{\ell+1})}{Pw(i_\ell)}\right) = E\left\{\prod_{\ell=0}^{N-2} \frac{w(i_\ell, i_{\ell+1})}{Pw(i_\ell)} E\left(\frac{w(i_{N-1}, i_N)}{Pw(i_{N-1})} \middle| N-1\right)\right\}$$

the conditional expectation being equal to 1 according to the definition of Pw ; whence

$$1 = E\left(\exp \sum_{\ell=0}^{N-1} \log \frac{w(i_\ell, i_{\ell+1})}{Pw(i_\ell)}\right) \geq \int \exp N \int \log \frac{w(i,j)}{Pw(i)} d\lambda$$

where λ represents the empirical law of the $(i_\ell, i_{\ell+1})$ and the first integral in the right terme considers realizations such that the i_ℓ's empirical law be near ν . Over this set, one can minorate the right integral

$$1 \geq P(N,\nu) \exp N\left\{\inf_\lambda \int \log \frac{w}{Pw} d\lambda\right\}$$

whence the result announced (with some continuity argument/ν).

In the independent case ($P_{ij} = \mu(j)$), we can verify that

$$h_p(\nu) = \int \log \frac{d\nu}{d\mu} d\nu = I(\nu|\mu)$$

information of ν with respect to μ , in the Kullback-Leibler sense. We can also rewrite $h_p(\nu)$ in the dependent case under the very intuitive form

$$h_p(\nu) = \inf_{\lambda} \int I\,(\lambda\,(.\mid i) \mid P\,(i,.))\,d\nu(i)$$

where the inf is taken over the λ having both margins equal to ν ;

or in another way

$$h_p(\nu) = \inf_{\lambda} \int \log \frac{\lambda(j\mid i)}{P_{i,j}}\, d\lambda\,(i,j) = \inf_{\lambda} I\,(\lambda \mid P)$$

We can see immediately what we should write in the n-markovian case !

The periodic case.

Let $I = Z^k$, $T = Z^k$ mod.park n net, $p_{i,j} = \pi\,(\underline{i},j-i)$.

We can then define $p_{\underline{i},\underline{j}} = \sum_{k=\underline{j}-\underline{i}} \pi\,(\underline{i},k)$

If $d\nu$ is a law over T x T having equal margins, we will write $P^{\nu}_{\underline{i},\underline{j}}$ the conditional law $\nu(\underline{j}\mid\underline{i})$. Starting from $d\nu$ and from $\lambda \in (\mathbb{R}^k)^*$, we define new transition probabilities over Z^k by

$$P^{\nu,\lambda}_{i,j} = P^{\nu}_{\underline{i},\underline{j}} \cdot \frac{P_{i,j}}{P_{\underline{i},\underline{j}}} \cdot \exp \lambda\,(j-i) \cdot C\,(\underline{i},\underline{j},\lambda)$$

The "constants" C must verify the normalisation conditions

$$1 = \sum_j P^{\nu,\lambda}_{i,j} = \sum_{\underline{j}} P^{\nu}_{\underline{i},\underline{j}}\, C\,(\underline{i},\underline{j},\lambda) \sum_{j\in\underline{j}} \frac{P_{ij}}{P_{\underline{i},\underline{j}}} \exp \lambda\,(j-i)$$

$\frac{P_{i,j}}{P_{\underline{i},\underline{j}}}$ "is" the j-i law conditionally to $\underline{i}$ and to $\underline{j}$; we write

$L_{\underline{i},\underline{j}}(\lambda)$ for its Laplace transform. If we take

$C(\underline{i},\underline{j},\lambda) = 1/L_{\underline{i},\underline{j}}(\ \ )$, the new probabilities $P_{\underline{i},\underline{j}}$ are still periodic and $P^{\nu,\lambda}_{\underline{i},\underline{j}}$ admits $d\nu$ as its joint invariant law. Let's calculate the Radon-Nikodym derivative along the trajectory $i_o,\ldots,i_N$ with a mean increment v.

$$\frac{1}{N}\log\frac{dP^{\nu,\lambda}}{dP} = \frac{1}{N}\Sigma\log\frac{P^{\nu}_{\underline{i}_\ell,\underline{i}_{\ell+1}}}{P_{\underline{i}_\ell,\underline{i}_{\ell+1}}} - \frac{1}{N}\Sigma\log L_{\underline{i}_\ell,\underline{i}_{\ell+1}}(\lambda)+\lambda v$$

For the $P^{\nu,\lambda}$-typical trajectories, we have

$$\frac{1}{N}\log\frac{dP^{\nu,\lambda}}{dP} \simeq I(\nu|\underline{P}) - L_\nu(\lambda) + \lambda v$$

where $L_\nu(\lambda) = \int \operatorname{Log} L_{\underline{i},\underline{j}}(\lambda)\, d\nu\, (\underline{i},\underline{j})$.

We now want to evaluate $P(\{$trajectories with a mean increment $v\}$: to get the formula we will prove the minoration.

1°) ν fixed, we choose λ so that v is the mean typical increment for $P^{\nu,\lambda}$, which is equivalent to minimizing $L_\nu(\lambda)-\lambda v$. Let $H_\nu(v)$ lenote the minimum.

°) We choose ν so that it minimizes $I(\nu|P) - H_\nu(v)$

$$h(v) = \sup_\nu\,[-I(\nu|P) + H_\nu(v)]$$

The studied probability is then "minorized" by $\exp - Nh(v)$.

applying a min-max argument, h(v) appears as the dual of

$$\mathcal{L}(\lambda) = \sup_\nu\,[L_\nu(\lambda) - I(\nu|P)] = \sup_\nu \int_{T\times T} \log\frac{L_{\underline{i},\underline{j}}(\lambda)\ P_{\underline{i},\underline{j}}}{P^{\nu}_{\underline{i},\underline{j}}}\, d\nu$$

If we write down the extremum conditions, we find that $\overset{\vee}{P}_{\underline{i},\underline{j}}$ must be of the form $\frac{1}{C(\lambda)} \frac{\varphi(\underline{j})}{\varphi(\underline{i})} L_{\underline{i},\underline{j}}(\lambda) P_{\underline{i},\underline{j}}$ and the normalization conditions give

$$\sum_{\underline{j}} L_{\underline{i},\underline{j}}(\lambda) P_{\underline{i},\underline{j}} \varphi(\underline{j}) = C(\lambda) \varphi(\underline{i})$$

Then $C(\lambda)$ is the eigenvalue associated to the positive eigenvector of $L_{\underline{i},\underline{j}}(\lambda) P_{\underline{i},\underline{j}}$ and $\mathcal{L}(\lambda) = \log C(\lambda)$.

§ 4. Fast simulations (continued).

Fast simulations with forced deformations.

Let's consider the last algorithm of § 1 : a batch of Poisson size and expectation μ is serialized by stopping when a failure and two successes are ascertained (or when we have a success from the start). The approximative analysis of this algorithm turns our interest to the random walk with independent increments τ_i whose law is that of $\theta - \frac{\mu}{\lambda} a$, where θ is the duration of the examination of the tree built from the original batch, and a the ratio of messages successfully sent.

It is difficult to obtain an analytical expression of the τ-law. If we determine it by simulation, we can deduce from it an estimation of the root of $\hat{\tau} = 1$, and then an idea of the time one has to wait for the random walk to get a fixed level ; this constitutes a lot of approximations with respect to the initial problem. We propose to accelerate the simulation of the algorithm itself, by modifying the two parameters that control its functioning : the output λ of message arrivals to the ring, and the probability π with which, when authorized, the individual counter of a terminal is incremented. When a run is made, it is very easy to compute the corresponding Radon-Nikodym derivative.

The abstraction of the situation is the following : ρ being a parameter, $f(x,\rho)\,dx$ is the jump law of a family of random walks. At ρ_o, the expectation of a jump is negative and we want to

B' A B estimate $Q^{\rho_o}(\Omega_B)$, where $\Omega_B = \{$trajectories stemming from A, exiting from BB' through B$\}$.

We estimate $P^{\rho_o}(\Omega_B)$ as an empirical mean of an N-sample, whose law is the image of dP^{ρ} through dP^{ρ_o}/dP^{ρ}.

Let's index the trajectories in Ω_B by the empirical law $\mu(x)dx$ of the incrementations following the trajectory, and let's write v the expectation of μ . For the trajectories $\omega \in \Omega_B^{\mu}$, we have :

$$\log \frac{dP^{\rho_o}}{dP^{\rho}}(\omega) \simeq \frac{B-A}{v} \int \log \frac{f(x,\rho_o)}{f(x,\rho)}\, \mu(x)\, dx$$

and
$$\log P^{\rho}(\Omega_B^{\mu}) \simeq \frac{B-A}{v} \int \log \frac{f(x,\rho)}{\mu(x)}\, \mu(x)\, dx$$

Asymptotically, we assimilate the variance of the estimator to the term given by the worst of the Ω_B^{μ} : whence a log-variance.

$$V_{\rho} = \sup_{\mu} \frac{1}{v} \int [2 \log f(x,\rho_o) - \log f(x,\rho) - \log \mu(x)]\, d\mu$$

For v fixed, the optimizing μ is

$$\mu(x) = \frac{f^2(x,\rho_o)}{f\,(x,\rho)} \exp\,(\alpha x + \beta)$$

with

$$\begin{cases} \exp\beta \displaystyle\int \frac{f^2(x,\rho_o)}{f(x,\rho)} \exp\alpha x\, dx = 1 \\ \exp\beta \displaystyle\int \frac{f^2(x,\rho_o)}{f(x,\rho)}\, x \exp\alpha x\, dx = v \end{cases}$$

Let's write $g(x) = \dfrac{f^2(x,\rho_o)}{f(x,\rho)}$. Optimizing with respect to α rather than v, we find

$$V_\rho = \sup_\alpha \left(-\alpha + \frac{\hat{g}_\rho(\alpha)\log\hat{g}_\rho(\alpha)}{\hat{g}'_\rho(\alpha)}\right)$$

We can easily check that the extremum is reached for α_ρ such that $\hat{g}_\rho(\alpha_\rho) = 1$, and is worth $-\alpha_\rho$.

To find the ρ minimizing V_ρ, we have to solve the system of equations

$$\begin{cases} \hat{g}_\rho(\alpha_\rho) = 1 \\ \dfrac{\partial\hat{g}_\rho}{\partial\rho}(\alpha_\rho) = 0 \end{cases}$$

which can also be rewritten in the following way :

$$\begin{cases} E^{\rho_o}\left(\dfrac{f(x,\rho_o)}{f(x,\rho)} \exp\alpha x\right) = 1 \\ E^{\rho_o}\left(\dfrac{f(x,\rho_o)}{f^2(x,\rho)}\dfrac{\partial}{\partial\rho} f(x,\rho)\exp\alpha x\right) = 0 \end{cases}$$

and this can be done by incrementing (α,ρ) by a stochastic gradient method.

Time before overflow for stabilized ALOHA.

Let's consider a message arriving to the system when the number of blocked messages is N_o, and let's evaluate the probability $P(N,T)$ that this message will have to wait a time T before a successfull transmission (it is then simple to ponder these probabilities by those a newcomer has of finding the system in the state N_o). We can define a new process associated to the "marked" message : the set of states is still $\mathbb{N}$, but there exists now a possibility of extinction ; at each moment the marked message is still in the system, we look at the numer $N(t)$ of blocked messages. This chain is still markovian.

As in § 2, we make space and time continuous ; we imbed our process into a family indexed by ε, $0 < \varepsilon \leq 1$, interpolating the $P_{i,\cdot}$ by $P_{x,\cdot}$ and defining the logarithm $L(x,\rho)$ of the Laplace transform of the law of increments starting from x(for $\varepsilon = 1$), as well as its dual function $h(x,v)$. An easy generalization of the quoted theorems shows that we obtain an evaluation of $P(N_o,T)$'s logarithm by minimizing over all the functions $\varphi : [0,T] \to \mathbb{R}$, who verify $\varphi(0)=N_o$, the action integral $\int_0^T h_{\varphi(t)}(\dot{\varphi}(t))\, dt$.

Let $\phi(N_o,T)$ be the minimum of the action. It verifies the equation of dynamic programming

$$\phi(x,t+dt) = \inf_{dx} \left[h\left(x, \frac{dx}{dt}\right) dt + \phi\,(x,dx,t)\right]$$

or

$$\frac{\partial\phi}{\partial t} = \inf_{v} \left[h\,(x,v) + \frac{\partial\phi}{\partial x}\,(x,t)\; v\right]$$

$$= L\,\left(x, \frac{\partial\phi}{\partial x}\,(x,t)\right)$$

and $\frac{\partial \phi}{\partial x}(x,t) = p(x,t)$ over the minimizing trajectory is still more interesting than v, for it represents the rate of exponential modification one must include in the law of increments to make absolutely typical those trajectories only typical conditionnaly they live a time > T.

P verifies the quasilinear equation :

$$\frac{\partial p}{\partial t} = L_x(x,p) + L_p(x,p)\,\frac{\partial p}{\partial x}$$

and the surface $p(x,t)$ is obtained by glueing the trajectories of the Hamilton-Jacobi system

$$\begin{cases} dx = v(x,p)\,dt \\ dp = -\,L_x\,dt \end{cases}$$

Since we are in dimension 1, the system's trajectories are the level curves of L. The action $\phi(N_o,T)$ is the integral of $-p\,dx + L(x,p)dt = h(x,v(x,p))$ along the trajectory such that $X(0)=N_o$ and $p(T)=0$. Practically, the level curves of L are parametered by the duration of travel between $x=N_o$ and $p=0$. We then have the actions $\phi(N_o,T)$, but also the probability changes $p(x,T)$ for a fast simulation. We can ask if this change should not simultaneously depend on the state attained by the simulated process and on the time spent since the beginning of the simulation, or only on this time (and this would have the advantage of being a one-to-one dependence). In fact, a change depending only of x allows us to assure the constancy of exp $L(x,p)$ during the runs, and this simplifies all computations, since it is a normalisation factor in the probability changes (we forgot it in § 2 since we worked with $L(x,p)=0$!).

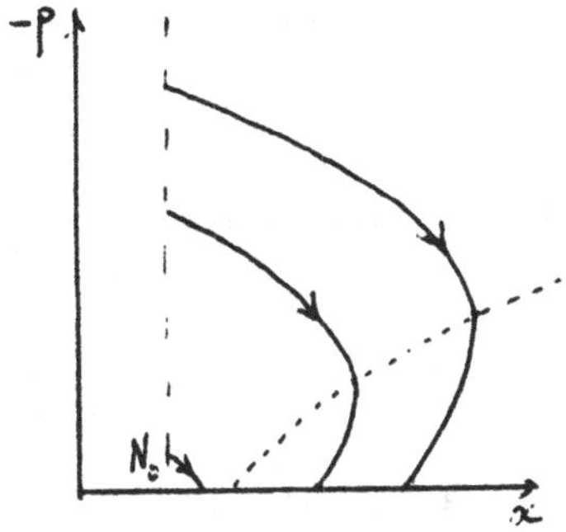

Bibliography :

[A-R] AZENCOTT-RUGET : Mélanges d'équations différentielles et grands écarts à la loi des grands nombres, Zeit. f. War. vol. 38, p. 1 (1977)

[B-G] BANH TRI AN and GELENBE : Near-optimal behavior of the packet switching broadcast channel, Rapport de recherche n° 13 (1978), L.R.I., Université Paris-Nord, Orsay.

[C-F-M] COTTRELL-FORT-MALGOUYRES : Preprint Université Paris-Sud et Proceedings of the International Conference on Information Sciences and Systems, Patras, Greece, July 9-13, 1979

[D-V] DONSKER-VARADHAN : Asymptotic evaluation of certain Markov process expectation for large time. Comm. Pure. Appl. Math. vol. 28, p. 1, vol. 29, p. 279 et vol. 29 p. 389.

[F-G-L] FAYOLLE-GELENBE-LABETOULLE : Stability and optimal control of the packet switching broadcast channel J.A.C.M., vol. 24, n° 3 (1977), p. 375

[G] GÄRTNER : On large deviations for the invariant measure Th. Prob. and Appl., vol. 22, p. 24 (1977)

[K] KHASMINSKII : A limit theorem for the solutions of differential equations with random right-hand sides Th. Prob. and Appl. vol. 11, p. 390 (1966)

[L] LUDWIG : Persistence of dynamical systems under random perturbations. Siam Review, vol. 17, n° 4, p. 605 (1975)

[M] MOLCHANOV : Diffusion processes and riemannian geometry. Russ. Math. Surveys, vol. 30, n° 1, p. 1 (1975)

[R] RUGET : Quelques occurences des grands écarts dans le littérature "électronique" dans le Séminaire de Statistique d'Orsay, 1977-1978, publ. by S.M.F. (Astérisque n° 68)

[V] VENTSEL : Rough limit theorems on large deviations for Markov stochastic processes. Th. Prob. and Appl., vol. 21, p. 227 et p. 499 (1976)

ASYMMETRIC BROADCAST CHANNELS

Andrea Sgarro
Istituto di Elettrotecnica e di Elettronica
e Istituto di Matematica
Università degli Studi di Trieste, Italy

1. Introduction

Broadcast channels have been introduced by Cover in a paper published in 1972 [1]; as an acknowledgement to the importance of this work he was granted the IEEE award in 1973. By now broadcast channels are firmly established as one of the most relevant channel networks for multi-terminal communication.

Broadcast channels attempt to model the situation of a radio or of a TV broadcaster where several pieces of information are to be sent from a single terminal input point to different terminal output points. For (relative) simplicity we shall consider only memoryless and stationary broadcast channels with two terminal output points.

The communication situation which we have in mind is described in fig. 1. There are three different message sets, $\mathcal{M}$, $\mathcal{M}'$ and $\mathcal{M}''$. Three independent messages, say i,j and k, are selected from the message sets and encoded using a single codeword, that is an n-length sequence belonging to the input alphabet of the broadcast channel. As a consequence of the channel noise two random sequences are received at the output terminals of the broadcast channel. The first user decodes the first output sequence to obtain estimates of i and j; the second user decodes the second output sequence to obtain estimates of i and k. Thus i represents common information meant for both users, while j and k represent private information meant for the first and the second user, respectively.

The following question is of interest: at which rates is reliable transmission possible in the communication situation described above? Obviously in trying to answer this question we shall assume the standpoint which is typical of Shannon theory: we shall ignore any complexity argument (time and money resources are supposed to be unlimited) and shall aim at asymptotic results (this implies that the transmission time is thought of as being infinite). It is well-known, however, that the highly idealized results of Shannon theory have nevertheless a normative value for the design of real-world communication systems.

A broadcast channel like that of fig. 1 is too difficult, at least so far, to be dealt with

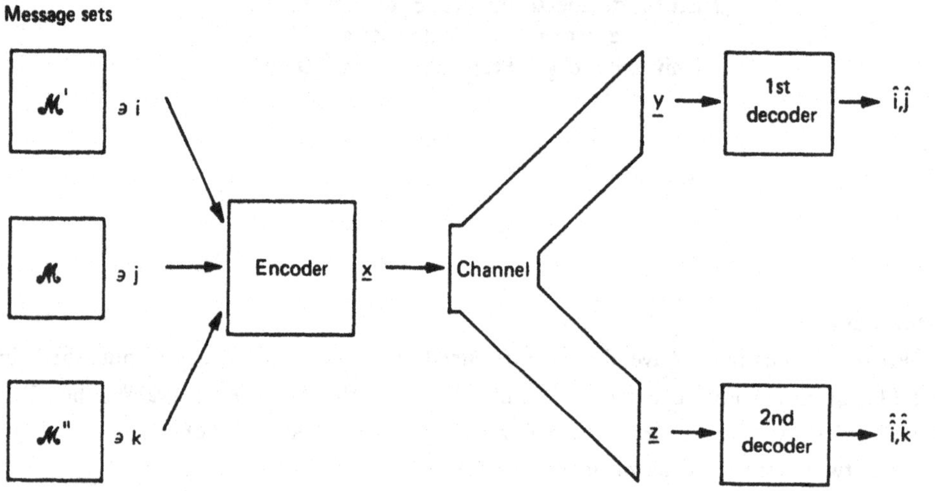

Fig. 1 : General broadcast channel

mathematically in its full generality. Thus situations have been considered where some sort of "degradation" is present. First [1,2], the degradation has been assumed in the transmission medium: the so called ***degraded broadcast channel*** is made up by two ordinary channels connected in series; (thus the output alphabet of the first channel coincides with the input alphabet of the second channel). The first user observes the output of the first channel while the second user observes the more noisy output of the second channel.

Later, however, Körner and Marton [3] shifted the degradation to the message sets, assuming that no private information is to be sent to the second user (message set $\mathcal{M}''$ has one element only, or, equivalently, its rate is zero; see fig. 2).

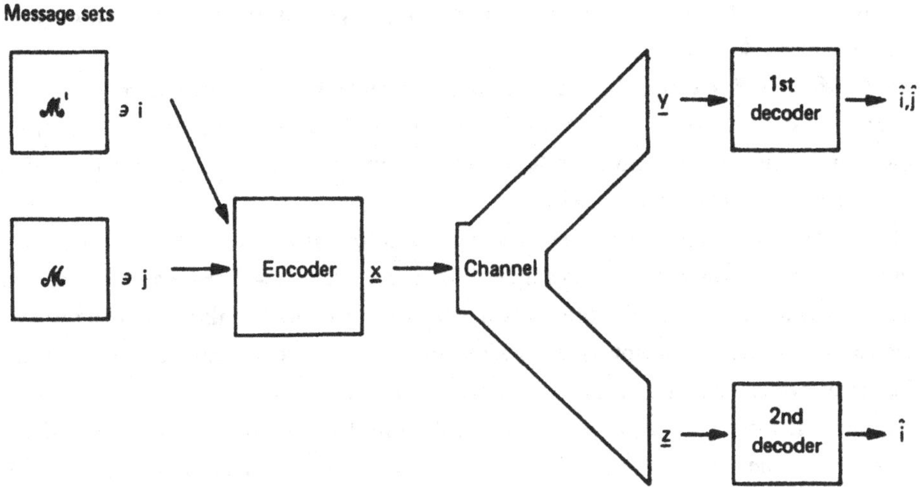

Fig. 2 : Asymmetric broadcast channel

The communication network of fig. 2 was originally called *broadcast channel with degraded message sets;* later Csiszár and Körner [4] turned to the handier name of *asymmetric broadcast channel.* We shall retain the latter terminology and the related acrostic ABC. We stress that the asymmetry of an ABC refers only to the message structure and not to the transmission medium as such, this being a quite arbitrary one-input and two-output channel.

A list of references containing original material on ABC's is given at the end of this paper [3,5,6]; an extensive presentation is to be found in the forthcoming book by Csiszár and Körner [4]. References [7] and [8] are useful surveys on multi-user communication.

In section 2 we give some preliminaries and notation; in section 3 we define ABC's with mathematical rigour; in section 4 we state some results on ABC's. More involved proofs are omitted in accordance with the expository character of this paper.

2. Preliminaries

Assume that $\mathcal{A}$ is an alphabet (that is a finite non-empty set) and that $\mathbf{a}$ is an n-length sequence of elements of $\mathcal{A}$. The *type* of $\mathbf{a}$ is the probability distribution (or p.d.) P over $\mathcal{A}$ defined by

$$P(a) \overset{\text{def}}{=} n^{-1} N(a|\mathbf{a}), \qquad a \in \mathcal{A},$$

where $N(a|\mathbf{a})$ is the number of occurrences of a in $\mathbf{a}$. Obviously a p.d. P on $\mathcal{A}$ is the type of some n-length sequence iff for every $a\in\mathcal{A}$ the number n P(a) is an integer.

In particular if $\mathcal{A}$ is a product-set, $\mathcal{A} = \mathcal{B}\times\mathcal{C}$, a type on $\mathcal{A}$ is called a *joint type;* it is a joint p.d. over $\mathcal{B}\times\mathcal{C}$.

If $\mathcal{A}$, $\mathcal{B}$ and $\mathcal{C}$ are alphabets and (A,B,C) is a three-dimensional random variable (or r.v.) over $\mathcal{A}\times\mathcal{B}\times\mathcal{C}$ H(A), H(B|A), I(A∧B) and I(A∧B|C) denote entropy, conditional entropy, mutual information and conditional mutual information, respectively (for standard Shannon theory concepts see, e.g., [4]).

Notice that "dummy" entropies and mutual informations can be associated with sequences and joint sequences. For example, if **b** and **c** are n-length sequences whose joint type Q is the p.d. of a r.v. (B,C), the mutual information I(B∧C) is also called the mutual information of **b** and **c**; obviously such a quantity has a merely combinatorial character and the use of probabilistic terms is simply a matter of convenience.

We say that three r.v.'s A, B and C form a Markov chain if I(A∧C|B) = 0; heuristically C can be seen as the output of two channels connected in series whose inputs are A and B, respectively.

If P is a p.d. over $\mathcal{A}$ and $V:\mathcal{A}\to\mathcal{B}$ and $W:\mathcal{A}\to\mathcal{B}$ are two stochastic matrices with input alphabet $\mathcal{A}$ and output alphabet $\mathcal{B}$, the divergence [4] of V from W weighted by P is defined as

$$D(V\|W|P) \overset{\text{def}}{=} \sum_{a\in\mathcal{A}}\sum_{b\in\mathcal{B}} P(a)\ V(b|a)\ \log\frac{V(b|a)}{W(b|a)}$$

(logs and exps will be taken to the base 2). The weighted divergence is non-negative; it is zero iff $V(\cdot|a) = W(\cdot|a)$ whenever P(a) is strictly positive ($V(\cdot|a)$ and $W(\cdot|a)$ are conditional p.d.'s over $\mathcal{B}$).

The cardinality of a set $\mathcal{A}$ is denoted by $|\mathcal{A}|$; $\mathcal{A}^n$ is the set of all n-length sequences of $\mathcal{A}$; $|r|^+$ denotes the positive part of the real number r, i.e. $|r|^+ = r$ if $r > 0$ and $|r|^+ = 0$ if $r < 0$.

3. Definitions

An *asymmetric broadcast channel,* or ABC, $W = (W_\mathcal{Y}, W_\mathcal{Z})$ is a pair of ordinary channels $W_\mathcal{Y}:\mathcal{X}\to\mathcal{Y}$, $W_\mathcal{Z}:\mathcal{X}\to\mathcal{Z}$ where $\mathcal{X}$, $\mathcal{Y}$ and $\mathcal{Z}$ are alphabets. We use the same symbols for channels and for the corresponding stochastic matrices. The n-th memoryless extension of W, W^n, is defined as the pair of channels $(W^n_\mathcal{Y}, W^n_\mathcal{Z})$, where $W^n_\mathcal{Y}$ and $W^n_\mathcal{Z}$ are the n-th memoryless extensions of $W_\mathcal{Y}$ and $W_\mathcal{Z}$.

Let $\mathcal{M}$ and $\mathcal{M}'$ be alphabets called the *common message set* and the *private message set,* respectively. A *code* for the n-th memoryless extension of the ABC $W = (W_\mathcal{Y}, W_\mathcal{Z})$ is a triple of mappings, (f, φ, ψ), $f: \mathcal{M} \times \mathcal{M}' \to \mathcal{X}^n$, $\varphi: \mathcal{Y}^n \to \mathcal{M} \times \mathcal{M}'$, $\psi: \mathcal{Z}^n \to \mathcal{M}$; f is called the *encoder,* φ and ψ are called the $\mathcal{Y}$-decoder and the $\mathcal{Z}$-decoder, respectively. The elements of $f(\mathcal{M} \times \mathcal{M}')$ are called *codewords;* n is called the *blocklength* of the code; the numbers $R = n^{-1} \log|\mathcal{M}|$ and $R' = n^{-1}\log|\mathcal{M}'|$ are called the *common rate* and the *private rate* of the code.

The event that the $\mathcal{Y}$-decoder (the $\mathcal{Z}$-decoder) is in error will be referred to as a $\mathcal{Y}$-error ($\mathcal{Z}$-error); the probabilities of $\mathcal{Y}$-error and of $\mathcal{Z}$-error for codeword $x_{i,j} \overset{\text{def}}{=} f(i,j)$, $i \in \mathcal{M}$, $j \in \mathcal{M}'$, are by definitions the numbers:

$$e^{(\mathcal{Y})}_{i,j} = W^n_{\mathcal{Y}}(\{y: \varphi(y) \neq (i,j)\} | x_{i,j}), \text{ and}$$

$$e^{(\mathcal{Z})}_{i,j} = W^n_{\mathcal{Z}}(\{z: \psi(z) \neq i\} | x_{i,j}),$$

respectively; $y \in \mathcal{Y}^n$, $z \in \mathcal{Z}^n$. The *maximum probabilities of $\mathcal{Y}$-error* and *$\mathcal{Z}$-error* are the numbers $e^{(\mathcal{Y})} = \max e^{(\mathcal{Y})}_{i,j}$ and $e^{(\mathcal{Z})} = e^{(\mathcal{Z})}_{i,j}$; the maximizations are carried out with respect to all values of the indices.

A pair of numbers $(E^{(\mathcal{Y})}, E^{(\mathcal{Z})})$ is called an *attainable pair of error exponents* at common rate R and private rate R′ for the ABC W if for any $\delta > 0$ and for all sufficiently large n there exist n-length codes of rate pair at least $(R-\delta, R'-\delta)$ having maximum probabilities of error $e^{(\mathcal{Y})}$ and $e^{(\mathcal{Z})}$ not exceeding $\exp\{-n[E^{(\mathcal{Y})}-\delta]\}$ and $\exp\{-n[E^{(\mathcal{Z})}-\delta]\}$, respectively, when used for the n-th memoryless extensions of W.

The *capacity region* of the ABC W is defined as the set of all real pairs (R,R') such that for any $\epsilon > 0$ any $\delta > 0$ and all sufficiently large n there exist n-length codes of rate pair at least $(R-\delta, R'-\delta)$ having maximum probabilities of error smaller than ϵ. Clearly, if there is a strictly positive attainable pair of error exponents at common rate R and private rate R′ for the ABC W, then (R, R') belongs to the capacity region of this ABC.

Given an ABC W and the positive real numbers R and R′, to every p.d. Q on $\mathcal{U} \times \mathcal{X}$, where $\mathcal{U}$ is an arbitrarily chosen dummy alphabet, we shall associate a pair of functions, $F^{(\mathcal{Y})}$ and $F^{(\mathcal{Z})}$, defined as follows:

$$F^{(\mathcal{Y})} = F^{(\mathcal{Y})}(W,R,R',Q) = \min[D(V_{\mathcal{Y}} \| W_{\mathcal{Y}} | Q) + \min(|I(U, X \wedge Y) - R - R'|^+, |I(X \wedge Y | U) - R'|^+)],$$

$$F^{(\mathcal{Z})} = F^{(\mathcal{Z})}(W,R,R',Q) = \min[D(V_{\mathcal{Z}} \| W_{\mathcal{Z}} | Q) + |I(U \wedge Z) - R|^+].$$

Here the first and the third minimum are taken with respect to all ABC's $V = (V_{\mathcal{Y}}, V_{\mathcal{Z}})$, $V_{\mathcal{Y}}: \mathcal{U} \times \mathcal{X} \to \mathcal{Y}$, $V_{\mathcal{Z}}: \mathcal{U} \times \mathcal{X} \to \mathcal{Z}$, with input alphabet $\mathcal{U} \times \mathcal{X}$. With a slight abuse of notation we have denoted by the symbol $W = (W_{\mathcal{Y}}, W_{\mathcal{Z}})$ also the ABC $\tilde{W} = (\tilde{W}_{\mathcal{Y}}, \tilde{W}_{\mathcal{Z}})$ with input alphabet $\mathcal{U} \times \mathcal{X}$ defined by:

$$\tilde{W}_{\mathcal{Y}}(y|u,x) = W_{\mathcal{Y}}(y|x),\ \tilde{W}_{\mathcal{Z}}(z|u,x) = W_{\mathcal{Z}}(z|x),\ u\in\mathcal{U},\ x\in\mathcal{X},\ y\in\mathcal{Y},\ z\in\mathcal{Z}.$$

The p.d. of the random triple (U,X,Y) is specified by Q and $V_{\mathcal{Y}}$, while the p.d. of (U,X,Z) is specified by Q and $V_{\mathcal{Z}}$.

Since $F^{(\mathcal{Y})}$ does not depend on $V_{\mathcal{Z}}$ and $F^{(\mathcal{Z})}$ does not depend on $V_{\mathcal{Y}}$, there is always a single ABC V which attains both minima appearing in the definitions of $F^{(\mathcal{Y})}$ and $F^{(\mathcal{Z})}$.

4. Results

Theorem 1 and 2 below are due to Körner and this author [6]; theorem 1 refers to fixed type codes, that is codes whose codewords are permutations of one another and therefore have all the same associated type.

Theorem 1 (Attainable Error Exponents for Fixed Type Codes). Let $\mathcal{U}$ be any alphabet. For every triple of positive real numbers R, R' and δ, for every $n \geq n_1(|\mathcal{U}|, |\mathcal{X}|, |\mathcal{Y}|, |\mathcal{Z}|, \delta)$, and for every joint type Q of sequences in $\mathcal{U}^n \times \mathcal{X}^n$ there exist codes of blocklength n and of rate pair at least $(R - \delta, R' - \delta)$ such that all codewords have the same type and for every ABC W the maximum probabilities of error satisfy the inequalities:

$$e^{(\mathcal{Y})} = e^{(\mathcal{Y})}(W) \leq \exp\{-n(F^{(\mathcal{Y})}(W,R,R',Q) - \delta)\},$$

$$e^{(\mathcal{Z})} = e^{(\mathcal{Z})}(W) \leq \exp\{-n(F^{(\mathcal{Z})}(W,R,R',Q) - \delta)\}.$$

The type of the codewords referred to in the theorem is the marginal type of Q over $\mathcal{X}$. For the proof of Theorem 1 see [6]; here we shall only mention a fact whose bearing will become clear later (cf. theorem 4). An inspection of the derivation of Theorem 1 in [6] shows that the codes which are used (or, if one prefers, the corresponding encoding and decoding rules) are *independent of the channel statistics.* In particular the decoding rules have a merely combinatorial (non-stochastic) character which deserves some mention (the original idea of this sort of combinatorial decoding goes back to Goppa [9]): when decoding one maximizes "combinatorial mutual informations" like those hinted at in section 2.

The following Theorem 2 is an easy corollary of Theorem 1:

Theorem 2 (Attainable Error Exponents) Let $\mathcal{U}$ be an alphabet and let Q be a p.d. over $\mathcal{U} \times \mathcal{X}$. Then $(F^{(\mathcal{Y})}(W,R,R',Q), F^{(\mathcal{Z})}(W,R,R',Q))$ is an attainable pair of error exponents for the ABC W.

Proof: Consider a sequence of types Q_n over $\mathcal{U} \times \mathcal{X}$ converging to Q (Q_n is the type of an n-length sequence). Theorem 2 follows from Theorem 1, since it can be easily shown that $F^{(\mathcal{Y})}$ and $F^{(\mathcal{Z})}$are continuous in Q uniformly in W.

Q.E.D.

A comment to this result is given after Theorem 3. Theorem 3 below is due to Körner and Marton [3]; however the direct part can be obtained as a corollary to Theorem 2.

Theorem 3. (Capacity Region of an ABC). The capacity region of the ABC $W = (W_{\mathcal{Y}}, W_{\mathcal{Z}})$ is the closure of the set $\mathcal{R}$ defined as follows:

$\mathcal{R} = \{(R, R')$: there exist an alphabet $\mathcal{U}, |\mathcal{U}| \leqslant |\mathcal{X}| + 2$, and a random quadruple (U,X,Y,Z) over $\mathcal{U} \times \mathcal{X} \times \mathcal{Y} \times \mathcal{Z}$ such that the conditional distributions of Y and Z given X are $W_{\mathcal{Y}}$ and $W_{\mathcal{Z}}$, respectively, U, X and (Y, Z) form a Markov chain in this order, and the following inequalities hold: $I(X \wedge Y) \geqslant R + R'$, $I(X \wedge Y | U) \geqslant U$, $I(U \wedge Z) \geqslant R\}$.

Proof: First notice that the condition $I(X \wedge Y) \geqslant R + R'$ can be easily written $I(UX \wedge Y) \geqslant R + R'$: in fact $I(UX \wedge Y) = I(X \wedge Y) + I(U \wedge Y|X) = I(X \wedge Y)$ because of markovity. An inspection of the positivity conditions for $F^{(\mathcal{Y})}$ and $F^{(\mathcal{Z})}$ shows that the closure of $\mathcal{R}$ is contained in the capacity region of the ABC W. We omit the proof of the converse result, that is that no rate pair outside the closure of $\mathcal{R}$ is attainable.

Q.E.D.

Comment 1: On looking back at Theorem 2 one sees retrospectively that the form of the error exponents is less haphazard than it might have seemed at first sight. In fact in the definitions of $F^{(\mathcal{Y})}$ and $F^{(\mathcal{Z})}$ one makes use of the "excess rates" $I(X \wedge Y) - R - R'$, $I(X \wedge Y | U) - R'$ and $I(U \wedge Z) - R$: cf. the form of the inequalities defining the capacity region of W. If these excess rates are "large", that is if one is well inside the capacity region, the error probabilities converge to zero at a large exponential speed.

Comment 2: The capacity region of a general (not asymmetric) broadcast channel is a set of real triples (R, R', R'') defined by extending in an obvious way the definition of the capacity region of an ABC. Theorem 3, which asserts that the latter equals the closure of $\mathcal{R}$ has some relevance also for the general case, since the projection of the capacity region of a general broadcast channel over the plane (R, R') coincides with the capacity region of the corresponding ABC. To see this it is enough to use the fact that if a triple (R, R', R'') is attainable then certainly also the "worse" triple (R − δ , R′ − δ, R″ − δ) is attainable (δ > 0); on the other hand, for R″ = 0 ($\mathcal{M}''$ has only one message) one obtains precisely the capacity region of the corresponding ABC.

Comment 3: Lastly, we shall make use of the observation which follows Theorem 1. A code is said to be *universal* for a class of ABC's with common input and output alphabet when it "performs well" for all the ABC's of that class. The following theorem, which is proved in the same way as Theorem 2 and which is due to the same authors, asserts the existence of universal codes with strictly positive error exponents for "large" classes of ABC's.

Theorem 4 **(Universally Attainable Error Exponents).** Let $\mathcal{U}$ be an alphabet and Q a p.d. over $\mathcal{U} \times \mathcal{X}$. Then $F^{(\mathcal{Y})}$ and $F^{(\mathcal{Z})}$ is a universally attainable pair of error exponents for the class of all ABC's with input alphabet $\mathcal{X}$ and output alphabets $\mathcal{Y}$ and $\mathcal{Z}$.

(We recall that a function $(F^{(\mathcal{Y})}, F^{(\mathcal{Z})})$, from $\mathcal{R}^+ \times \mathcal{R}^+ \times \mathcal{W}$ into $\mathcal{R} \times \mathcal{R}$ is called a universally attainable pair of error exponents for the class $\mathcal{W}$ of all ABC's with input alphabet $\mathcal{X}$ and output alphabets $\mathcal{Y}$ and $\mathcal{Z}$ if, for every R > 0, R′ > 0, δ > 0, and for all sufficiently large n, $n \geq n_2(|\mathcal{X}|, |\mathcal{Y}|, |\mathcal{Z}|, \delta)$, there exist n-length codes (f_n, φ_n, ψ_n) of rate pair at least (R − δ, R′ − δ) such that, if W is an ABC of the above class and if $e_n^{(\mathcal{Y})}(W)$ and $e_n^{(\mathcal{Z})}(W)$ are the maximum probabilities of error of (f_n, φ_n, ψ_n) for the n-th memoryless extension of W, one has $e_n^{(\mathcal{Y})}(W) \leq \exp\{-n[F^{(\mathcal{Y})}(R,R',W)]\}$ and $e_n^{(\mathcal{Z})}(W) \leq \exp\{-n[F^{(\mathcal{Z})}(R,R',W)]\}$; above $\mathcal{R}$ $(\mathcal{R}^+)$ is the set of all (strictly) positive real numbers).

REFERENCES

[1] T.M. Cover: "Broadcast Channels", IEEE-IT 18, 2-14, 1972.

[2] P.P. Bergmans: "Random Coding Theorems for Broadcast Channels with Degraded Components", IEEE-IT 19, 197-207, 1973.

[3] J. Körner, K. Marton: "General Broadcast Channels with Degraded Message Sets", IEEE-IT 23, 60-64, 1977.

[4] I. Csiszár, J. Körner: "Information Theory. Coding Theorems for Discrete Memoryless Systems", Academic Press, forthcoming.

[5] G.S. Poltyrev: lecture at the Hungarian-Soviet-Czechoslovak Seminar on Information Theory, Tsahksador, 10-17 Sept. 1978.

[6] J. Körner, A. Sgarro: "Universally Attainable Error Exponents for Broadcast Channels with Degraded Message Sets", IEEE-IT, to appear.

[7] E.C. Van der Meulen: "A Survey of Multi-way Channels in Information Theory: 1961-1976", IEEE-IT 23, 1-37, 1977.

[8] J. Körner: "Some Methods in Multi-user Communication: a Tutorial Survey", in "Information Theory. New Trends and Open Problems", edited by G. Longo, CISM Courses and Lect ures No. 219, Springer-Verlag, 1975.

[9] V.D. Goppa: "Nonprobabilistic Mutual Information without Memory" (in Russian, Problems Control Inf. Th. 4, 97-102, 1975.

A LINEAR PROGRAMMING BOUND FOR CODES IN A 2-ACCESS BINARY ERASURE CHANNEL

Henk C.A. van Tilborg
Department of Mathematics
Eindhoven University of Technology
Eindhoven
the Netherlands

ABSTRACT. Two counting arguments are used to derive a system of linear inequalities which give rise to an upperbound on the size of a code for a 2-access binary erasure channel.
For uniquely decodable codes this bound reduces to a purely combinatorial proof of a result by Liao. Examples of this bound are given for some codes with minimum distance 4.

I. INTRODUCTION

Consider the 2-access communication system as depicted in figure 1. During a message interval, the two messages $\underline{m}_1$ and $\underline{m}_2$ from the two sources are independently encoded according to two binary codes C_1 and C_2 of the same length n. The two codewords are combined by the channel into a single vector $\underline{r}$ with symbols from {0,1,2}. The single decoder at the receiver decodes $\underline{r}$ into two codewords, one in C_1 and one in C_2, for the two users.

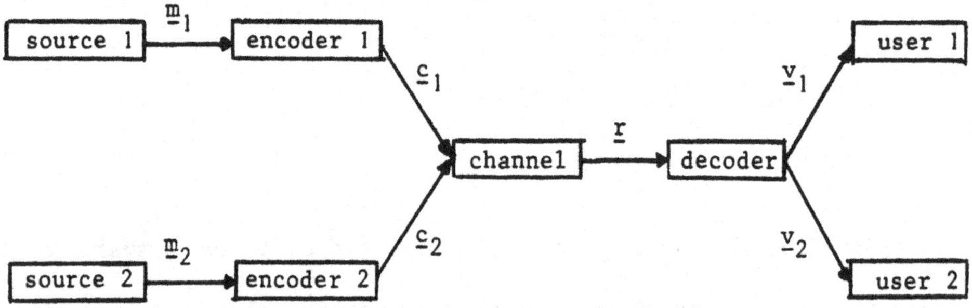

Fig. 1 Two-access communication system

The channel, which will be a 2-user adder channel, is depicted in figure 2. We shall say that a single error transmission error has occured if any of the following transitions took place:

i) from (0,0) or (1,1) to 1,

ii) from (0,1) or (1,0) to 0 or 2.

Two transmission errors have occured if one of the following transitions took place:

i) from (0,0) to 2,

ii) from (1,1) to 0.

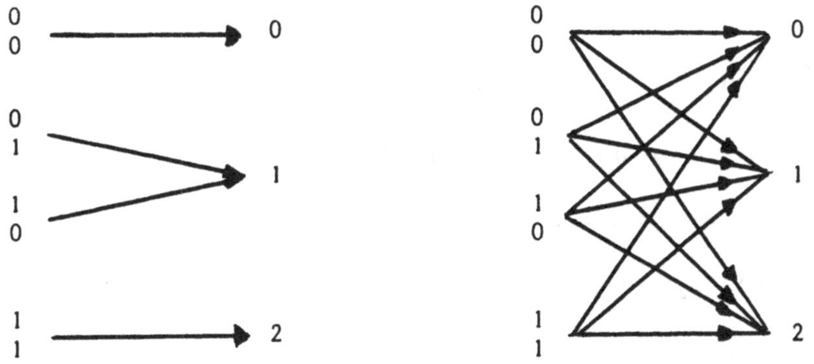

Fig. 2

Noiseless 2-user adder channel — Noisy 2-user adder channel

Various studies of this channel have been made, see ref. 1-6. For further background we refer the reader to Kasami and Lin[3]. In Kasami and Lin[4] lower and upper bounds on the achievable rates of the codes C_1 and C_2 are derived, assuming that C_1 is linear (or systematic).

In this paper we shall show that by solving a linear programming problem, one can find an upper bound on the sum of the rates of the two codes without any additional assumptions.

II. DEFINITIONS AND LEMMAS

We adopt the definitions of Kasami and Lin[4]. Let $V_n = \{0,1\}^n$ and $W_n = \{0,1,2\}^n$. Let $\phi : V_n \times V_n \to W_n$ be defined by

$$\phi(\underline{u},\underline{v}) = \underline{u}+\underline{v}\ , \quad \underline{u} \in V_n,\ \underline{v} \in V_n, \tag{2.1}$$

where the addition is componentwise over the reals.

Let C_1 and C_2 be two codes in V_n. Then we shall call $\phi(C_1,C_2)$

defined by

$$\phi(C_1,C_2) = \{\phi(\underline{c}_1,\underline{c}_2) \mid \underline{c}_1\in C_1,\ \underline{c}_2\in C_2\} \tag{2.2}$$

a code for the 2-access binary erasure channel.

Definition 2.1. A code $\phi(C_1,C_2)$ is called uniquely decodable if for all $\underline{c}_1,\underline{c}_1' \in C_1$ and $\underline{c}_2,\underline{c}_2' \in C_2$, one has that $\phi(\underline{c}_1,\underline{c}_2) = \phi(\underline{c}_1',\underline{c}_2')$ implies that $\underline{c}_1 = \underline{c}_1'$ and $\underline{c}_2 = \underline{c}_2'$.
It follows from (2.2) and this definition that a code is uniquely decodable if and only if $|C_1|\times|C_2| = |\phi(C_1,C_2)|$, where $|\cdot|$ denotes set cardinality.

Example: For n = 2 take $C_1 = \{(0,0),(1,1)\}$ and $C_2 = \{(0,0),(0,1),(1,0)\}$. Then, $\phi(C_1,C_2)$ is uniquely decodable, as follows readily from the following table.

TABLE I

A uniquely decodable code

$C_1 \backslash C_2$	0 0	0 1	1 0
0 0	0 0	0 1	1 0
1 1	1 1	1 2	2 1

We shall now define a distance d_L in W_n. For $\underline{x} \in W_n$ and $\underline{y} \in W_n$, define $d_L(\underline{x},\underline{y})$ by

$$d_L(\underline{x},\underline{y}) = \sum_{i=1}^{n} |x_i-y_i| \quad , \tag{2.3}$$

where the subtraction again takes place over the real numbers and $|x_i-y_i|$ denotes the absolute value of x_i-y_i.

We refer to this distance as the L-distance of $\underline{x}$ and $\underline{y}$ (to distinguish it from the normal Hamming distance, denoted by $d_H(\underline{x},\underline{y})$).

Definition 2.2. The minimum L-distance d_L of a code $\phi(C_1,C_2)$ in W_n is defined by

$$d_L = \min\{d_L(\phi(\underline{c}_1,\underline{c}_2),\phi(\underline{c}_1',\underline{c}_2')) \mid (\underline{c}_1,\underline{c}_2) \neq (\underline{c}_1',\underline{c}_2'),$$

$$\underline{c}_i,\underline{c}_i' \in C_i,\ i = 1,2\}.$$

Clearly a code $\phi(C_1,C_2)$ is uniquely decodable if and only if its minimum L-distance is greater than or equal to one.

Let [x] denote the greatest integer less than or equal to x. Suppose that $\phi(C_1,C_2)$ is a code in W_n with minimum L-distance $d_L \geq 1$. Let the number of transmission errors, as defined in the introduction, not exceed $[(d_L-1)/2]$. Then the received word $\underline{r}$ will be closer to the transmitted word $\phi(\underline{c}_1,\underline{c}_2)$ than to any other word in $\phi(C_1,C_2)$. In other words this code is able to correct up to $[(d_L-1)/2]$ errors. We now quote a result from Kasami and Lin[3].

Lemma 2.3. Let $\phi(C_1,C_2)$ be a code with minimum L-distance d_L. Then the Hamming distance of C_1 as well of C_2 is greater than or equal to d_L.

Proof. Let $\underline{c}_1$ and $\underline{c}_1'$ be 2 different codewords in C_1 and let $\underline{c}_2 \in C_2$. Then it follows from (2.3) that the Hamming distance of $\underline{c}_1$ and $\underline{c}_1'$ equals $d_L(\phi(\underline{c}_1,\underline{c}_2),\ \phi(\underline{c}_1',\underline{c}_2'))$, which is greater than or equal to d_L. The same argument holds for C_2. □

III. AN UPPER BOUND

Throughout this paragraph, $\phi(C_1,C_2)$ will denote a code of length n and minimum L-distance d_L with $d_L \geq 1$.

We shall describe a method for determining an upper bound for:

$$M(n,d) := \max\{|\phi(C_1,C_2)| \mid \phi(C_1,C_2) \text{ is a code of length n and minimum L-distance } d_L\} \quad (3.1)$$

We start with some definitions.

$$W_{n,k} := \{\underline{u} \in W_n \mid \underline{u} \text{ has exactly k coordinates equal to 1}\}, \quad (3.2)$$

$$D_k := \{\underline{u} \in \phi(C_1,C_2) \mid \underline{u} \in W_{n,k}\}, \quad (3.3)$$

$$d_k := |D_k|. \quad (3.4)$$

Clearly

$$\sum_{k=0}^{n} d_k = |\phi(C_1,C_2)| = |C_1| \times |C_2|. \quad (3.5)$$

We shall derive two sets of linear inequalities on the numbers d_k. The upper bound on M(n,d) will be the maximum value of $\sum_{k=0}^{n} d_k$ subject to these inequalities. For $0 \leq i,k \leq n$ we define:

$$F_i(k) = \{(\underline{u},\underline{c}) \mid \underline{u} \in W_{n,k},\ \underline{c} \in \phi(C_1,C_2),\ d_L(\underline{u},\underline{c}) = i\}, \quad (3.6)$$

$$f_i(k) = |F_i(k)|, \quad (3.7)$$

$$m(k,i) = \max_{\underline{x} \in W_{n,k}} |\{\underline{c} \in \phi(C_1,C_2) \mid d_L(\underline{x},\underline{c}) = i\}|. \quad (3.8)$$

Clearly any $\underline{u} \in W_{n,k}$ contributes at most once to $\bigcup_{i=0}^{e_L} F_i(k)$, where $e_L := [(d_L-1)/2]$.

Lemma 3.1. For $0 \le i, k \le n$

$$f_i(k) = \sum_{2p+q+r=i} \binom{k+r-q}{r} \binom{n-k-r+q}{p} \binom{n-k-r+q-p}{q} 2^r d_{k+r-q}.$$

Proof. Let $u \in W_{n,\ell}$. Change r ones of $\underline{u}$ into a zero or a two. Change q zero or two coordinates into a two resp. a zero. Obviously this can be done in

$$\binom{l}{r}\binom{n-l}{p}\binom{n-l-p}{q} 2^r$$

different ways. The words obtained in this way are in $W_{n,\ell-r+q}$ and have L-distance 2p+q+r to $\underline{u}$. Now set ℓ-r+q to k, set 2p+q+r to i and let $\underline{u}$ run through D_{k+r-q} for all possible (p,q,r). □

Example

$$f_0(k) = d_k$$

$$f_1(k) = 2(k+1)d_{k+1} + (n-k+1)d_{k-1}$$

$$f_2(k) = 4\binom{k+2}{2}d_{k+2} + 2k(n-k)d_k + (n-k)d_k + \binom{n-k+2}{2}d_{k-2}.$$

Since a vector $y \in W_{n,k}$ can contribute more than once to $F_i(k)$ for $i > e_L$ we need an upper bound on m(k,i).

Lemma 3.2. Let $d_L = 2e_L+2$. Then $m(k,e_L+1) \le \dfrac{2n}{e_L+1}$.

Proof. Let $\underline{u} \in W_{n,k}$ and let $\underline{c}^{(1)},\dots,c^{(m)}$ be all vectors in $\phi(C_1,C_2)$ at L-distance e_L+1 from $\underline{u}$. Since $d_L = 2(e_L+1)$, the L-distance of $\underline{c}^{(i)}$ and $\underline{c}^{(j)}$, for $i \ne j$, is exactly $2(e_L+1)$. More precisely for all $1 \le \ell \le n$

and $i \neq j$, one has

$$d_L(c_\ell^{(i)}, c_\ell^{(j)}) = d_L(c_\ell^{(i)}, x_\ell) + d_L(x_\ell, c_\ell^{(j)}).$$

This implies that for all $1 \leq \ell \leq n$ either $c_\ell^{(i)} = x_\ell$ or $x_\ell = c_\ell^{(j)}$ or finally $x_\ell = 1$, while $c_\ell^{(i)}$ and $c_\ell^{(j)}$ are different and not equal to 1. Hence at each 1 coordinate (non-1 coordinate) of $\underline{x}$ at most 2 (resp. 1) codewords $\underline{c}^{(i)}$ may have a different entry from $\underline{x}$.

Now for each $\underline{c}^{(i)}$, $1 \leq i \leq m$, let

$$a_i(1) = |\{1 \leq \ell \leq n| \quad x_\ell = 1, \quad c_\ell^{(i)} \neq 1\}|,$$

$$a_i(2) = |\{1 \leq \ell \leq n| \quad x_\ell \neq 1, \quad c_\ell^{(i)} = 1\}|,$$

$$a_i(3) = |\{1 \leq \ell \leq n| \quad (x_\ell, c_\ell^{(i)}) = (0,2) \text{ or } (2,0)\}|.$$

It follows from the observations above that

$$\sum_{i=1}^{m} a_i(1) \leq 2k,$$

$$\sum_{i=1}^{m} (a_i(2) + a_i(3) \leq n-k,$$

while clearly $a_i(1) + a_i(2) + 2a_i(3) = d_L(\underline{c}^{(i)}, \underline{x}) = e_L$. Hence

$$m(e_L+1) = \sum_{i=1}^{m} (a_i(1)+a_i(2)+2a_i(3)) \leq$$

$$\sum_{i=1}^{m} (a_i(1)+2a_i(2)+2a_i(3)) \leq 2k+2(n-k) = 2n$$

i.e. $m \leq \dfrac{2n}{e_L+1}$.

□

We are now able to formulate our first set of inequalities on the numbers d_k, $0 \leq k \leq n$.

Theorem 3.3.

$$\sum_{i=0}^{e_L} f_i(k) \le \binom{n}{k} 2^{n-k}, \qquad \text{if } d_L = 2e_L+1 \tag{3.9}$$

$$\sum_{i=0}^{e_L} f_i(k) + \frac{e_L+1}{2n} f_{e_L+1}(k) \le \binom{n}{k} 2^{n-k}, \text{if } d_L = 2e_L+2 \tag{3.10}$$

where the $f_i(k)$ can be expressed in terms of the numbers d_k by lemma 3.1.

Proof. Since $|W_{n,k}| = 2^{n-k}$, (3.9) follows from the observation that its left hand side equals the number of words $\underline{u} \in W_{n,k}$ at L-distance at most e_L to $\phi(C_1,C_2)$. Similarly the left hand side of (3.10) is, by lemma 3.2, a lower bound for the number of words $\underline{u} \in W_{n,k}$ at L-distance at most e_L+1 to $\phi(C_1,C_2)$.

□

The inequalities 3.9 or 3.10 seem to be rather strong for large values of k (in terms of n) but very weak for small values of k. Our next system of inequalities will have the opposite behaviour.

Theorem 3.4. Let $A(n,d_H) := \max\{|C| \mid C \text{ is a code in } V_n \text{ with minimum Hamming distance } d_H\}$. Then

$$\sum_{i=k}^{\ell} \binom{\ell}{k} d_\ell \le \binom{n}{k} A(k,[\frac{d_L+1-2(\ell-k)}{2}]), \tag{3.11}$$

where $0 \le k \le \ell \le n$ and $2(\ell-k) \le d_L-1$.

Proof. It is sufficient to prove for all $1 \le i_1 < \ldots < i_k \le n$ that

$$|\{\underline{c} \in \bigcup_{i=k}^{\ell} D_i \mid c_{i_1} = \cdots = c_{i_k} = 1\}| \le A(k,[\frac{d_L+1-2(\ell-k)}{2}]), \tag{3.12}$$

since summation of (3.11) over all possible $i_1,\ldots,i_k$ yields (3.11). W.l.o.g. we can take $(i_1,\ldots,i_k) = (1,2,\ldots k)$. Let $(\underline{c}_1,\underline{c}_2) \neq (\underline{c}_1',\underline{c}_2')$, both in (C_1,C_2), be such that $\underline{c} = \phi(\underline{c}_1,\underline{c}_2) \in W_{n,p}$, $\underline{c}' = \phi(\underline{c}_1',\underline{c}_2') \in W_{n,q}$, where $k \le p,q \le \ell$ and where $c_i = c_i' = 1$, $1 \le i \le k$. We can arrange the coordinates such that for some r,

$$\longleftarrow k \longrightarrow \longleftarrow r \longrightarrow \longleftarrow p-k-r \longrightarrow \longleftarrow q-k-r \longrightarrow \longleftarrow n-p-q+k+r \longrightarrow$$

$$\underline{c} = (1,1,\ldots,1,1,1,\ldots,1,1,1,\ldots\ldots,1\ , c_{p+1},\ldots,c_{p+q-k-r},c_{p+q-k-r+1},\ldots,c_n)$$

$$\underline{c}' = (1,1,\ldots,1,1,1,\ldots,1,c_{k+r-1}',\ldots,c_p',1,1,\ldots\ldots,1 \qquad ,c_{p+q-k-r+1}',\ldots,c_n')$$

where $c_i \neq 1$ for $p+1 \le i \le n$ and $c_i' \neq 1$ for $k+r-1 \le i \le p$ and $p+q-k+r+1 \le i \le n$. Let $\bar{x}$ denote $1-x$. Then

$$\underline{c}_1 = (c_1,\ldots,c_k,c_{k+1},\ldots,c_{k+r},c_{k+r+1},\ldots,c_p,c_{p+1},\ldots,c_{p+q-k-r},c_{p+q-k-r+1},\ldots,c_n)$$

$$\underline{c}_2 = (\bar{c}_1,\ldots,\bar{c}_k,\bar{c}_{k+1},\ldots,\bar{c}_{k+r},\bar{c}_{k+r+1},\ldots,\bar{c}_p,c_{p+1},\ldots,c_{p+q-k-r},c_{p+q-k-r+1},\ldots,c_n)$$

$$\underline{c}_1' = (c_1',\ldots,c_k',c_{k+1}',\ldots,c_{k+r}',c_{k+r+1}',\ldots,c_p',\ c_{p+1}',\ldots,c_{p+q-k-r}',c_{p+q-k-r+1}',\ldots,c_n')$$

$$\underline{c}_2' = (\bar{c}_1',\ldots,\bar{c}_k',\bar{c}_{k+1}',\ldots,\bar{c}_{k+r}',c_{k+r+1}',\ldots,c_p',\bar{c}_{p+1}',\ldots,\bar{c}_{p+q-k-r}',c_{p+q-k-r+1}',\ldots,c_n').$$

We can now determine $\phi(\underline{c}_1,\underline{c}_2)$ and $\phi(\underline{c}_1',\underline{c}_2)$ and compute their L-distance.

$$d_L((\phi(\underline{c}_1,\underline{c}_2')),(\phi(c_1',c_2))_i = \begin{cases} d_L(\phi(c_i,\bar{c}_i'),\phi(c_i',\bar{c}_i)) = 2d_H(c_i c_i'), & 1 \le i \le k+r, \\ d_L(\phi(c_i,c_i').\phi(c_i',\bar{c}_i)) = 1 \ . \ , & k+r+1 \le i \le p, \\ d_L(\phi(c_i,\bar{c}_i'),\phi(c_i',c_i)) = 1 \ . \ , & p+1 \le i \le p+q-k-r, \\ d_L(\phi(c_i,c_i'),\phi(c_i',c_i)) = 0, & p+q-k-r+1 \le i \le n. \end{cases}$$

Hence the L-distance of $\phi(\underline{c}_1,\underline{c}_2')$ and $\phi(\underline{c}_1',\underline{c}_2)$ on the last n-k positions is at most $2r+(p-k-r)+(q-k-r) = (p-k)-(q-k) \leq 2(\ell-k)$. So the L-distance on the first k positions is at least $d_L-2(\ell-k)$. Consequently the Hamming distance of $\underline{c}_1$ and $\underline{c}_1'$ on the first k positions is at least $(d_L-2(\ell-k))/2$ i.e. at least $[(d_L+1-2(\ell-k))/2]$. Hence the number of codewords in $\phi(C_1,C_2)$ with the first k coordinates equal to 1 is at most $A(k,[(d_L+1-2(\ell-k))/2]$. □

Theorems 3.3 and 3.4 together imply our main theorem.

Theorem 3.5. Let $\phi(C_1,C_2)$ be a code of length n with minimum L-distance $d_L \geq 1$. Then $\phi(C_1,C_2) = |C_1|\times|C_2|$ is less than or equal to the maximum value of $\sum_{k=0}^{n} d_k$ subject to the inequalities (3.9) and (3.11) for d_L is odd and (3.10) and (3.11) for d_L is even.

IV. UNIQUELY DECODABLE CODES

In the case of $d_L = 1$, (3.9) reduces to $d_k \leq \binom{n}{k} 2^{n-k}$ and (3.11) reduces to $d_k \leq \binom{n}{k} 2^k$, since $A(k,1) = 2^k$. So the next theorem is a direct application of theorem 3.5.

Theorem 4.1. Let $\phi(C_1,C_2)$ be a uniquely decodable code of length n. Then

$$|C_1|\times|C_2| \leq \begin{cases} 2\sum_{k=0}^{m} \binom{2m+1}{k} 2^k & , \quad n = 2m+1, \\ 2\sum_{k=0}^{m} \binom{2m+2}{k} 2^k + \binom{2m+2}{m+1} 2^{m+1} & , \quad n = 2m+2. \end{cases}$$

In Liao[6] it is shown that the sum of the rates of the codes C_1 and C_2 cannot exceed 3/2. Theorem 4.1 now gives a combinatorial proof of the same result since.

$$2\sum_{k=0}^{m}\binom{2m+1}{k}2^k \le 2^{m+1}\sum_{k=0}^{m}\binom{2m+1}{k} = 2^{3m+1} < 2^{3n/2},$$

and

$$2\sum_{k=0}^{m}\binom{2m+2}{k}2^k + \binom{2m+2}{m+1}2^{m+1} \le 2^{m+1}\sum_{k=0}^{m+1}\binom{2m+2}{k} \le 2^{3m+3} = 2^{3n/2}.$$

We shall now analyze the asymptotic behavior of the sums on the right in Theorem 4.1. Since

$$\sum_{k=0}^{m}\binom{2m+1}{k}2^k = \sum_{\ell=0}^{m}\binom{2m+1}{m-\ell}2^{m-\ell} = 2^m\binom{2m+1}{m}\sum_{\ell=0}^{m}\frac{m(m-1)\cdots(m-\ell+1)}{(m+2)(m+3)\cdots(m+\ell+1)}\cdot\frac{1}{2^\ell},$$

we have on one hand

$$2\sum_{k=0}^{m}\binom{2m+1}{k}2^k \le 2^{m+1}\binom{2m+1}{m}\sum_{\ell=0}^{m}\frac{1}{2^\ell} \le 2^{m+2}\binom{2m+1}{m}; \tag{4.1}$$

while on the other hand

$$2\sum_{k=0}^{m}\binom{2m+1}{k}2^k = 2\sum_{\ell=0}^{m}\binom{2m+1}{\ell}2^{m-\ell} \ge$$

$$2^{m+1}\binom{2m+1}{m}\sum_{\ell=0}^{[m^{\frac{1}{4}}]}\left(1-\frac{2}{m+2}\right)\left(1-\frac{4}{m+3}\right)\cdots\left(1-\frac{2\ell}{m+\ell+1}\right)\frac{1}{2^\ell} \ge$$

$$2^{m+1}\binom{2m+1}{m}\sum_{\ell=0}^{[m^{\frac{1}{4}}]}\left(1-\frac{\ell(\ell+1)}{m+2}\right)\frac{1}{2^\ell} \ge$$

$$2^{m+2}\binom{2m+1}{m}\left(1-\frac{m^{\frac{1}{4}}(m^{\frac{1}{4}}+1)}{m+2}\right)\left(2-2^{-[m^{\frac{1}{4}}]}\right) =$$

$$2^{m+2}\binom{2m+1}{m}\left(1+\mathcal{O}\left(\frac{1}{\sqrt{m}}\right)\right), \quad (m\to\infty). \tag{4.2}$$

The inequalities (4.1) and (4.2) and their analogon for $n = 2m+2$ prove the following theorem.

Theorem 4.2. Let $\phi(C_1, C_2)$ be a uniquely decodable code of length n. Then

$$|C_1| \times |C_2| \leq \begin{cases} \frac{2^{3n+3}}{\sqrt{\pi m}} \left(1 + \mathcal{O}\left(\frac{1}{\sqrt{m}}\right)\right) & n = 2m+1, \quad (m \to \infty), \\ \frac{3.2^{3m+3}}{\sqrt{\pi m}} \left(1 + \mathcal{O}\left(\frac{1}{\sqrt{m}}\right)\right) & n = 2m+2, \quad (m \to \infty). \end{cases}$$

Proof. Apply Stirling's formula. □

By Theorem 4.1, $M(2,1) \leq 6$ and $M(3,1) \leq 14$. In fact $M(2,1) = 6$ and $M(3,1) = 14$ (to see this, take for C_1 the all-zero and the all-one vectors and take for C_2 the whole vector space minus the all-one vector. By taking the direct sum of these codes, we have $M(2n,1) \geq 6^n$, $M(2n+1,1) \geq 14 \cdot 6^{n-1}$.

V. SOME EXAMPLES FOR $d_L = 4$.

It follows from lemma 2.3 that $M(n,d_L) \leq (A(n,d_L))^2$. Since $A(4,4) = A(5,4) = 2$ and $A(6,4)$ one can thus conclude that $M(4,4) \leq 4$, $M(5,4) \leq 4$ and $M(6,4) \leq 4$. As a matter of fact these three inequalities hold with equality as follows from the following two tables

TABLE II
M(4,4)=4

		C_2 1 1 0 0	0 0 1 1
C_1	0 0 0 0	1 1 0 0	0 0 1 1
	1 1 1 1	2 2 1 1	1 1 2 2

TABLE III

m(6,4)=16

	C_2 101010	100101	011001	010110
C_1 000000	101010	100101	011001	010110
111100	212110	211201	122101	121210
110011	211021	210112	121012	120121
001111	102121	101212	012112	011221

We shall now investigate the cases n = 7,8 and 16, mainly because in these cases we can compare our bounds with some known constructions (see Kasami and Lin[3]).

Since $A(7,4) = 2^3$, $A(8,4) = 2^4$ and $A(16,4) = 2^{11}$, lemma 2.3 yields $M(7,4) \le 2^6$, $M(8,4) \le 2^8$ and $M(16,4) \le 2^{22}$.

Since $A(k,2) = A(k-1,1) = 2^{k-1}$, for $k \ge 1$, one gets from (3.11) by taking $\ell = k$ and $\ell = k+1$.

$$D_k \le \binom{n}{k} 2^{k-1} \quad , \ 1 \le k \le n, \tag{5.1}$$

$$D_k + (k+1)D_{k+1} \le \binom{n}{k} 2^k \quad , \ 0 \le k \le n-1. \tag{5.2}$$

For $k \ge 1$ (5.2) implies (5.1), since $k+1 \ge 2$. So we can replace (3.11) by (5.2) in theorem 3.5. For n = 7 and 8 we could do the calculations by hand. For n = 16 a computer program was written by Mr. A. Rubin. The results were as follows:

$m(7,4) \le 53 \approx 2^{5.73}$,

$m(8,4) \le 145 \approx 2^{7.18}$,

$m(16,4) \le 266\,825 \approx 2^{18.03}$.

In Kasami and Lin[3] one can find the following two codes

i) $n = 8,\ d_L = 4,\quad |C_1| = 8,\quad |C_2| = 9$,

ii) $n = 16,\ d_L = 4\quad |C_1| = 32,\quad |C_2| = 1177$.

Since $M(n-1,d_L) \geq \frac{1}{4}M(n,d_L)$ (by shortening C_1 and C_2 in the appropriate way), we get the following values:

$M(4,4) = 4$, $20 \leq M(7,4) \leq 53$

$M(5,4) = 4$, $72 \leq M(8,4) \leq 145$

$M(6,4) = 16$, $37644 \leq M(16,4) \leq 266\,825$.

REFERENCES.

1. Ahlswede, R., "Multi-way communication channels", in Proc. 2nd Int. Symp. Information Transmission, Hungarian Press, Tsahkadsor, Armenia, U.S.S.R.

2. Gaarder, N.T. and J.K. Wolf, "The capacity region of a multiple-access discrete memoryless channel can increase with feedback", IEEE Trans. Inform. Theory, vol. IT-21, pp.100-102, Jan. 1975.

3. Kasami, T. and S. Lin, "Coding for a multiple-access channel", IEEE Trans. Inform. Theory, vol. IT-22, pp.129-137, Mar. 1976.

4. Kasami, T. and S. Lin, "Bounds on the achievable rates of block coding for a memoryless multiple-access channel", IEEE Trans. on Inform. Theory, vol. IT-24, pp.187-197, Mar. 1978.

5. Kasami, T. and S. Lin, "Decoding of δ-decodable codes for a multiple-access channel", IEEE Trans. on Inform. Theory, vol. IT24, pp. 633-635, Sep. 1978.

6. Liao, H.H.J., "Multiple access channels", Ph.D. Thesis, Dep. Elec. Eng., University of Hawaii, Honolulu, 1972.

LIST OF CONTRIBUTORS

Toby BERGER, School of Electrical Engineering, Cornell University, Ithaca, NY 14853.

E. GELENBE, Laboratoire de Recherche en Informatique, Université de Paris-Sud, 91405 Orsay, France.

János KÖRNER, Mathematical Institute of the Hungarian Academy of Sciences.

James L. MASSEY, Professor of System Science, University of California, Los Angeles, Los Angeles CA, 90024.

Michael B. PURSLEY, Coordinated Science Laboratory, University of Illinois, Urbana, Illinois 61801 USA.

Gabriel RUGET, Mathématiques, Université Paris-Sud (Centre d'Orsay) 91405 Orsay (France)

Andrea SGARRO, Istituto di Elettrotecnica e di Elettronica e Istituto di Matematica, Università degli Studi di Trieste, Italy.

Henk C.A. van TILBORG, Department of Mathematics, Eindhoven University of Technology, Eindhoven, the Netherlands.